PRAISE FOR *THE DAWN OF A MINDFUL UNIVERSE*

"An extraordinary book. Marcelo Gleiser has brought together cosmology, environmentalism, and spirituality in a personal and poetic call to arms that is nothing short of breathtaking. Most of the time I was smiling and nodding as I read it, and occasionally I was moved to tears."

—William Egginton, author of *The Rigor of Angels*, *The Splintering of the American Mind*, and *The Man Who Invented Fiction*

"Marcelo Gleiser argues that the only hope we have of addressing the current environmental crisis lies in rethinking our relationship to history and to the entire cosmos. *The Dawn of a Mindful Universe* is a work of great honesty and daring. Its message couldn't be more alarming, yet it is ultimately optimistic."

—Elizabeth Kolbert, Pulitzer Prize–winning author of *The Sixth Extinction* and *Under a White Sky*

ALSO BY MARCELO GLEISER

The Island of Knowledge: The Limits of Science and the Search for Meaning

Great Minds Don't Think Alike: Debates on Consciousness, Reality, Intelligence, Faith, Time, AI, Immortality, and the Human

The Simple Beauty of the Unexpected: A Natural Philosopher's Quest for Trout and the Meaning of Everything

The Prophet and the Astronomer: A Scientific Journey to the End of Time

The Dancing Universe: From Creation Myths to the Big Bang

A Tear at the Edge of Creation: A Radical New Vision for Life in an Imperfect Universe

THE DAWN OF A MINDFUL UNIVERSE

A MANIFESTO FOR HUMANITY'S FUTURE

MARCELO GLEISER

An Imprint of HarperCollinsPublishers

THE DAWN OF A MINDFUL UNIVERSE. Copyright © 2023 by Marcelo Gleiser. All rights reserved. Printed in the United States of America. No part of this book may be used or reproduced in any manner whatsoever without written permission except in the case of brief quotations embodied in critical articles and reviews. For information, address HarperCollins Publishers, 195 Broadway, New York, NY 10007.

HarperCollins books may be purchased for educational, business, or sales promotional use. For information, please email the Special Markets Department at SPsales@harpercollins.com.

FIRST EDITION

Designed by Kyle O'Brien

Library of Congress Cataloging-in-Publication Data is available upon request.

ISBN 978-0-06-305687-9

23 24 25 26 27 LBC 5 4 3 2 1

To Earth, the planet that makes our story possible

CONTENTS

PROLOGUE...1

PART I | WORLDS IMAGINED

CHAPTER ONE Copernicus Is Dead! Long Live Copernicanism!...11

CHAPTER TWO Dreaming Up the Cosmos...20

PART II | WORLDS DISCOVERED

CHAPTER THREE The Desacralization of Nature...71

CHAPTER FOUR The Search for Other Worlds...89

CHAPTER FIVE Life on Other Worlds...135

PART III | THE UNIVERSE AWAKENS

CHAPTER SIX The Mystery of Life...149

CHAPTER SEVEN Lessons from a Living Planet...173

PART IV | THE MINDFUL COSMOS

CHAPTER EIGHT Biocentrism...191

CHAPTER NINE A Manifesto for Humanity's Future...197

EPILOGUE The Resacralization of Nature...208

ACKNOWLEDGMENTS...211

NOTES...213

INDEX...231

PROLOGUE

> No witchcraft, no enemy action had silenced
> the rebirth of new life in this stricken world.
> The people had done it themselves.
>
> —Rachel Carson, *Silent Spring*

The Universe has a history only because we are here to tell it. Through our diligence and ingenuity, we have pieced together the main chapters of the long saga that began with the Big Bang 13.8 billion years ago. This story unfolds in the vastness of space, narrating the drama of matter dancing to the tune of attractive and repulsive forces, shaping into ever more complex structures that became atoms, stars, galaxies, planets, life, us. The way we tell a story makes a difference. And it's time to retell the story of who we are under a new mindset. This book is about life on Earth, its cosmic relevance, about humanity's moral mandate to rise above our past to reshape our collective future. I write it with a sense of urgency and hope.

The advent of life changed everything. Life is matter with purpose, with an urge to exist. On this planet, the only known abode of life, a species unlike any other emerged some three hundred thousand years ago. *Homo sapiens*, us. What made us different from our bipedal ancestors was an enlarged frontal cortex that endowed us with an expansive capacity for symbolic thinking, combined with the manual dexterity to fashion raw materials into tools. We learned to control fire; we invented languages and learned to survive in groups, forging bonds of love and trust. We learned to tell stories, to inspire, to educate, and to caution. Through words and art, we recorded the past and imagined the future.

But let us not overromanticize our past. Tribes fought tribes for land and power, just as we still do, inflicting much suffering and bloodshed, just as we still do. Only now we are more efficient at it. Despite the violence, and differently from today, our ancestors kept one bond as holy: the bond to the land. To them, Nature was a sacred realm, and spirits animated the world and its mysteries. For millennia, Indigenous cultures across the globe have honored this tradition, revering the interconnectedness of all life. They have always known that we are not above Nature but part of the life collective, that our existence is fragile, contingent on powers beyond our control. They have always known that planet and life are one. We, their modern descendants, have forgotten all of this.

Our success at surviving changed us. From hunter-gatherers we clustered in agrarian societies, growing in numbers, bending Nature to our needs, finding ever more efficient ways to feed our physical hunger for food and our psychological hunger for power. We took ownership of the land, pushing the gods to the heavens. Earth lost its enchantment and became objectified, a thing to use with disregard, a place of change and decay, of sinful humans and wild beasts. What was once sacred became open to plundering. The life collective was ruptured, and creatures lost their right to exist.

Science amplified our success a thousandfold. The iron we mined, the coal we extracted, the gas and oil we burned, and the laws of mechanics and thermodynamics drove the industrial machinery that shaped the modern world. We grew in numbers and in need of resources, mining deeper, digging deeper, sucking the entrails of the planet for the fuel we so desperately wanted. The skies grew gray, the waters grew turbid, the air grew foul, forests were razed, and animals were systematically killed for food and pleasure.

A profound shift in perspective happened after Nicolaus Copernicus proposed in 1543 that, contrary to what everyone thought until then, the Earth was not the center of everything, but a mere planet orbiting the Sun, like all the others. This change in outlook was as confounding as it was revolutionary, and it's known as *Copernicanism*. Over the ensuing centuries, it framed how we see Earth's place in the Universe. Our planet, modern astronomy tells us, is just a rock orbiting a common star, an insignificant world floating among trillions of others in the empty vastness of space. Regrettably, Copernicanism evolved from the correct description of our planet's position in the solar system into a statement about our cosmic insignificance. Even life lost its magic, as we placed ourselves above animals, believing humans to be more like gods than beasts. The materialistic view of our planet and of life triumphed, portraying nonliving and living matter as machinelike things made of atoms, an amoral view that does not care for the environment or for the life collective.

Even if awe drives the creativity of many scientists, it is science's unavoidable alliance with the machinery of progress that shapes its quest. In its need for pragmatism, the orthodox scientific worldview killed Nature's spirit.

As we now enter a new age for humanity, the digital age, many aspire to take this view to its logical and horrifying conclusion—the final rejection of our bodies, our bonds to the life collective—and turn our essence into information stored in digital devices, the end of humanity as we know it. The technical unfeasibility and immorality of such mad transhuman dreams are beside the point. The point is the growing belief that this is our destiny, that to be codified into bits of information is our path to self-transcendence. As more people realize every day, this is the worldview that must change, the view that holds the planet and life as worthless and expendable, that places humans above Nature, that believes that our technological prowess alone will secure the future of civilization. If this view must go, the question is, How? How can we change our collective mindset? How can we remove the shadow that hovers over humanity, a shadow of our own making that threatens our collective future?

The premise of this book is that we need to reinvent ourselves as a species. The body of this book is my attempt to explain how. This is no utopian fantasy. We need to rewrite the story of who we are. To keep on going as if nothing is happening is simply not sustainable. Worse, it's delusional and suicidal.

And what is this new story for humankind? Science can guide us, if we shift our perspective. This new story connects us with life and the Universe, placing us as part of a biosphere that exists only because the Universe has allowed us to be, a story that expresses the interconnectedness of all that exists, what Buddhist master Thich Nhat Hanh called inter-being.

If, in the unfolding of time, the Universe or our galaxy had evolved differently, if a single event had changed in the history of life on Earth, we wouldn't be here. The asteroids and comets that came crashing down from the skies and other cataclysmic disasters over billions of years framed the course of evolution, shaping the creatures that could survive in the changing environment. The post-Copernican narrative I present here promotes the preciousness of our planet, its rarity as the only known cosmic jewel sparkling with a vibrant biosphere. This story links us and all life to a single bacterium that lived some three billion years ago on primal Earth, a story that looks at the multitude of worlds in our galaxy to point out how rare life is and, much more so, intelligent life capable of creating technologies to explore its cosmic origins. Instead of the depressing "the more we know about the Universe, the less relevant we become," I argue that "the more we know about the Universe, the more relevant we become." *We*, here, encompass the whole of planet Earth, a planet blessed with life and with a species capable of knowing its own history.

When our ancestors began to tell stories about human origins, about our search for meaning, the Universe gained a voice it had never had. Even if other voices are out there—and we don't know either way—they will never tell the cosmic story as we do. Their story will never be our story. As we will see, we are the only humans in the Universe, and the way we see the Universe is ours only. Without us, the Universe wouldn't know it exists. This is the story we tell. No other intelligence will tell it the same way. Through our voice, time became enfolded in memory and space became the stage where matter performed wonders. Through our voice, atoms formed and reshaped into stars and into living creatures. Through our voice, the Universe began to sing its song of Creation. *God's voice?*

The realization that we have a cosmic role, that we are interconnected with all that exists, that we are codependent with the life collective on this planet, has the power to reshape our destiny. There is no "us" without the biosphere. And without us, the biosphere doesn't know it exists, doesn't have a voice. The mechanistic narrative that has shaped our past must give way to a post-Copernican biocentric narrative, to a renewal of our spiritual bond to the land and to life, to a reenchantment of the planet. We can be successful only if we see each other as a single tribe, the human tribe, as we embrace our collective future with our hearts fired by the conviction that, together, we can be more than we have been.

※

I wrote this book as a call to action. The world is changing, and faster than we had imagined. Climate models have warned us for decades of what was to come, and now we are witnessing the consequences of our ways: tropical species migrating northward; storms becoming more powerful every year; the ongoing Sixth Extinction, an accelerated loss of biodiversity due to our encroachment on natural habitats and predatory hunting; the Anthropocene, the proposed name for the current geological era marked by our destructive presence; cities across the planet choking under a smoke-heavy sky; devastating droughts across the globe. The list goes on and on. Denying the effect of climate change on the planet is like denying that we age as time passes. But this is not another doomsday book, another warning about the inevitable darkness that awaits us. We have plenty of excellent ones already.[1]

Given the level of our inaction and inability to change, it should be clear that scare tactics are not working. They don't work because the effects of climate change are gradual and spread out, and they fluctuate because of the complexity of how geophysical systems couple with the biosphere. They don't work because change requires sacrifice at different fronts, from the individual to the corporate, asking for a profound realignment of the way we relate to the natural world. Climate change requires people to think in the long term, a no-no in a society geared toward short-term gain. What, then, would motivate such a profound change, given that we have consistently devalued the natural world for centuries? Why should people care about Nature, when they believe they are above it, that Nature is conveniently there for us to do with as we please?

To enact change, we first need to transform our collective mindset. We have to reassess our place in Nature and our impact on this planet and its biosphere. To enact change, we need to start by telling a new story. This book is my attempt to propose a post-Copernican worldview that realigns humanity with the natural world. The core principle behind this new worldview is *biocentrism*, the idea that a living planet is a sacred realm that deserves respect and veneration. I argue that this realization comes with a new moral imperative that, if followed, will redefine our collective future and ensure the longevity of our project of civilization.

Chapter 9, "A Manifesto for Humanity's Future," summarizes the book's message and presents the necessary steps toward securing the future of our civilization in a thriving biosphere. Whether or not those steps will be sufficient to change our current disastrous course is up to us, up to the whole human tribe. Even if most of us have done little to contribute to our current predicament, we will all share its consequences. When united with a single purpose, we do have the power to enact change. This is how revolutions have happened in the past. This is how we can do it again.

Not a revolution but a mindful evolution — Dec 25 12pm 2023

PART I

WORLDS IMAGINED

CHAPTER ONE

COPERNICUS IS DEAD! LONG LIVE COPERNICANISM!

> At rest, however, in the middle of everything is the sun. For in this most beautiful temple, who would place this lamp in another or better position than that from which it can light up the whole thing at the same time?
>
> —Nicolaus Copernicus, *On the Revolutions of the Heavenly Spheres*

Paralyzed by a stroke, the old astronomer lay in bed in helpless solitude. Mustering all his might, he propped his head up to catch a glimpse of the night sky through the window. His eyes wandered, like the planets, scanning the dark landscape of the heavens, the only place he felt at home. The stars would come and go, slowly rotating out of sight, until they came back the next night—sparks of light affixed to the dome of heaven. "What fools we have been," he mumbled to himself, "thinking that everything is as our eyes tell us." *The atomic eye — The philosopher*

Every morning, he eagerly waited for the visit of Tiedemann

Giese, a church canon like him and his only lifelong friend. Copernicus would stare at the door, anticipating the sound of Giese climbing the stairs. At ten o'clock, the old canon opened the door without knocking. "These stairs are going to be the death of me!" he said, gasping for air. Copernicus smiled as best he could and gestured for his friend to prop him up in bed. He pointed a trembling finger to the carefully wrapped package Giese had in his hand. "Yes, this is it, old man, your book is finally ready! It took you thirty years to write it, and it shows. It weighs a ton!"

It was Copernicus's life's work, packed within two covers: *On the Revolutions of the Heavenly Spheres*. The world would finally know what he thought about the Church's wrongheaded worldview. And the Church was not alone. The Babylonians, the Egyptians, the Greeks, the Romans—everyone had been wrong for thousands of years. The sole exception was the Greek Aristarchus. As early as 250 BCE he saw Earth for what it is, a planet circling the Sun. But no one listened. Aristotle's world system placing the Earth at the center of the Cosmos with the Moon, Sun, planets, and stars revolving around it was so simple and compelling that it had held all minds under a spell. Until now. Copernicus's book would fix this. He even dedicated it to Pope Paul III, expressing his hope that scripture and astronomy need not clash. God made the heavens. That was beyond dispute. To Copernicus, to be an astronomer was to worship the Lord's Creation. Only the stars could lift the minds of men closer to God's. The Holy Book, however, was no blueprint of Creation. It was not supposed to describe the Cosmos in any detail. Even if souls and planets are wanderers, they wander in different universes.

And now it was up to him, Nicolaus Copernicus, to reveal to the world the true message from the stars: that the Earth moves around the Sun just like Mars, Jupiter, and all other planets; that the Earth rotates about itself in twenty-four hours, the duration of a day; that the Moon is the only celestial object that goes around the Earth; and, finally, that all planets revolve around the Sun in circular orbits. Their arrangement follows the time it takes for each to complete a trip around the Sun: Mercury, three months; Venus, eight; Earth, one year; Mars, two; Jupiter, twelve years; and Saturn, the last, twenty-nine. Time is the secret to celestial harmony. This is the true message from the stars.

Giese sat by his friend's side and carefully unwrapped the package. As he flipped the cover open to look at the front pages, he noticed something unusual—a new preface, unsigned, not in the original manuscript. Johann Petreius, the publisher from Nuremberg, certainly wouldn't have written it. Georg Joachim Rheticus, Copernicus's only pupil, venerated his master's every word and wouldn't dare alter anything without permission. Who then?

Giese unsuccessfully tried to hide the extraneous page from his friend. But the trembling finger pointed right at it. Giese cleared his throat and read:

> *There have already been widespread reports about the novel hypotheses of this work, which declares that the Earth moves whereas the Sun is at rest in the center of the Universe. Hence certain scholars, I have no doubt, are deeply offended and believe that the liberal arts, which were established long ago on a sound basis, should not be thrown into confusion* ... [1]

"Maybe I should skip this," said Giese, a cold wave flooding his stomach. "Sounds like some fluff to start things up. Maybe Rheticus wrote this as a surprise to you?" The finger kept pointing resolutely at the page. Giese knew there was no going back. "Fine! Here it goes then." He skipped down a few sentences:

> *In this science there are some other no less important absurdities, which need not be set forth at the moment. For this art, it is quite clear, is completely and absolutely ignorant of the causes of the apparent nonuniform motions. And if any causes are devised by the imagination, as indeed very many are, they are not put forward to convince anyone that they are true, but merely to provide a reliable basis for computation.*

"'Merely to provide a reliable basis for computation?' This is rubbish!" Giese blurted. "This idiot is saying that your system of the world is a fantasy!" Torn by guilt, he looked at his ailing friend. He and Rheticus were the ones who had pushed Copernicus into writing the book, against his will. *He was right*, Giese thought. *The world is not ready for this kind of knowledge.*

Words locked inside his mind, Copernicus stared silently at the open window. A tear rolled down from his left eye, the one he could still open. The finger kept pointing at the book. Giese knew he had to finish:

> *Therefore alongside the ancient hypotheses, which are no more probable, let us permit these new hypotheses also to become known, especially since they are admirable as well as simple and bring with them a huge treasure of very skillful*

observations. So far as hypotheses are concerned, let no one expect anything certain from astronomy, which cannot furnish it, lest he accept as the truth ideas conceived for another purpose, and depart from this study a greater fool than when he entered it. Farewell.

Giese shook his head in disbelief. "I will take this to court tomorrow! We will fix this outrageous violation of your work! Who would do this? The coward didn't even sign it."

※

In a late 1543 letter to Rheticus, Giese registered the tragedy: "Copernicus only saw his completed book at the last moment, on the day he died." Giese tried to avenge his friend, but the courts would have none of it. For decades, most scholars who read the book believed that Copernicus had authored the anonymous preface claiming that the sun-centered model was simply a mathematical tool, not the true order of the planets. The perpetrator of this farce was, in fact, Lutheran theologian Andreas Osiander, who had corresponded with Copernicus over the years, arguing against his ideas. While at that point the Vatican was silent with respect to the arrangement of the heavens, Martin Luther had publicly criticized Copernicus's early ideas about a sun-centered Cosmos, calling him a "foolish astrologer."

Few episodes in the history of science are more significant or more dramatic. Rheticus, a Lutheran himself, had been put in charge of overseeing the manuscript's publication in Nuremberg. However, he had to flee town before the book was ready, apparently because of accusations of homosexuality. Osiander, a respected theologian, must have been the only local person Rheticus knew who was knowledgeable enough to take on the task. And so he did, adding his own preface and changing the original title from *On the Revolutions of the World Spheres* to *On the Revolutions of the Heavenly Spheres*, probably to get his point across right away. No world like ours revolves—only the spheres carrying planets up in the heavens. Osiander's message was clear: Copernicus's sun-centered Cosmos was but a fancy geometrical model with rotating spheres carrying planets around the sky, good for calculating their future locations and nothing else. The model had nothing to do with reality. Only fools would think otherwise.

Fifty years would pass before someone realized that Copernicus couldn't have written this preface. Apparently, the sleuth was German astronomer Johannes Kepler, who unmasked Osiander's fakery in 1609, if not earlier. Osiander's preface is crossed out with a big red *X* in Kepler's copy of the book.

In *The Book Nobody Read: Chasing the Revolutions of Nicolaus Copernicus*, astronomer and historian of science Owen Gingerich reconstructed the fate of the extant copies of Copernicus's book, uncovering them from oblivion in monasteries and private libraries or as they were passed from owner to owner across Europe. His conclusion: very few people cared about Copernicus's book, other than as a guide to predict the positions of the planets and stars, useful for astrology and navigation. Gingerich's book title sums it up. The publication of *On the Revolutions* didn't spark a revolution or even a noticeable reaction. Instead, the profound shift from an Earth-centered to a Sun-centered worldview would simmer in the background for a while, coming to a full boil only in the early 1600s, mostly because of Galileo Galilei in Italy and Johannes Kepler in Central Europe. These two trailblazing thinkers cared much more about the truth they could read from Nature than about faith-based dogmatic pronouncements. To both, careful observation and data analysis took precedence over Church authority. Measuring was how one read the book of Nature.

After thousands of years as the center of the Cosmos, Earth was pushed aside to join the other known planets. No centrality, no divine importance, no special mission or reason to exist. Just a wandering world, like so many, circling the Sun. This shift in the cosmic order changed history. When Earth lost its central status, so did humankind and the creatures of this world. This loss of centrality caused material and spiritual confusion. Before, with Earth as the center of Creation, things made sense. A rock falls to the ground to go back to where it belongs. Made of flesh and blood—earthy and liquid stuff—human beings tread along the dirt while their immaterial souls aspire to ascend to Heaven, to rejoin God. The vertical order of the physical Cosmos mirrored the Christian Cosmos, as depicted in Dante's *Divine Comedy*. The physical and the religious formed a cohesive wholeness. Now, new questions abounded. Why do things fall to the ground? Shouldn't they fall to the Sun if it is indeed the center of everything? Where is Heaven? Do other planets have living creatures? If so, are they also part of God's Creation?

I often wonder whether Copernicus knew his work would bring about such a profound change in worldview. I suspect he did, but we will never know. With the removal of Earth from the center of everything, what was unique to our planet became possible elsewhere. Especially life. In the 1580s, the bombastic Italian friar Giordano Bruno, perhaps the first outspoken Copernican, speculated that every star was a Sun surrounded by worlds, many of them inhabited, like ours. That being the case, and with other humans out there, sinners would abound in the Cosmos. Did they have a redeemer too? Was he the same Christ of our world? During the early 1600s, Kepler wrote a story called *Somnium*, in which a voyager travels to the Moon. Upon arrival, the explorer meets all sorts of creatures, bizarre mutations from what exists here, cave-dwelling, shadow-crawling, each with its own strange adaptations to an alien environment, foretelling to a degree what would become Darwin's theory of evolution some two and a half centuries later.

Once Earth is seen as just another planet among many, and given that the laws of physics and chemistry are the same across the Universe—and we now know that they are—life becomes, at least hypothetically, a cosmic imperative. Earth is no longer a special world. There should be multitudes of Earthlike worlds in our galaxy and likely as many in the billions of other galaxies spread across the Universe. If that's the case, if there are indeed many Earthlike planets out there, why not life? This, in a nutshell, is the essence of the Copernican worldview: there is nothing special about our planet; it is just a rocky world spinning around an ordinary star in the empty vastness of the Cosmos. This view is central to the profound identity crisis that threatens the future of our species and of many of the creatures with which we share this planet.

Worldviews change. They have changed in the past and will continue to change, as long as we care to learn more about the

Universe and our place in it. We are primed for change. Almost five centuries after the death of Copernicus, we have a new message from the stars: Copernicanism must go. It's time for a post-Copernican worldview to take over, informed by science and by a confluence of cross-cultural narratives that, taken together, can spark a profound change for humanity, a change with the power to reorient our collective future. For this change to take place, the current narrative must switch from one where Earth is a typical planet to one that celebrates the rarity of our planet and the life it shelters. We are the only species we know capable of this realization. After almost four billion years of evolution, our emergence on this rare planet marked the dawn of a new cosmic age: the cognitive age, the age of a mindful Universe. To know this and to internalize its meaning is to gain a new sense of collective purpose that asks us to reorient our relation to our living planet from one of abuse and neglect to one of reverence and gratitude. We are life capable of telling its own story. The story of what comes next, of the future of our collective project of civilization, is ours to make.

CHAPTER TWO

DREAMING UP THE COSMOS

> Worlds on worlds are rolling ever
> From creation to decay,
> Like the bubbles on a river
> Sparkling, bursting, borne away.
>
> —Percy Bysshe Shelley, *Hellas*

FROM MYTHS TO MODELS

Curiosity drives the imagination and rescues life from the triviality of sameness. This has always been true, but rarely with the explosive intensity of the philosophers who lived approximately from the sixth to the fourth centuries BCE in ancient Greece, collectively known as the *pre-Socratics*. The name implies that they lived before or around the time of Socrates, the Athenian philosopher who proposed that dialogue was the path to learning and mutual understanding. Up to then, gods had been the default explanation for why and how things happened the way they did—from natural disasters to victorious battles, from famine to times of plenty. The Sun crossed the sky every day from east to west as the god Helios in his fiery flying carriage. In Hinduism, Shiva danced the Cosmos into existence, animating matter and shaping it in its myriad forms before destroying his creation in endlessly recurring cycles. Such mythic explanations of natural phenomena are common to cultures across the globe, ancient and current. Poetic narratives that offer some control over powers that vastly overwhelm our own, they attempt to create a sense of order in a complex and often unpredictable world. Such myths transform natural events into stories people tell to make sense of what's seemingly beyond comprehension. Myths are grounding narratives that define cultural values, unifying ideas shared by a group. Their power lies not in their being right or wrong, but in their being believed. Myths translate Nature into words, humanizing the awesomeness of reality, bridging the concrete and the unknowable.

A famous Greek myth tells the story of Prometheus, the Titan who stole fire from the gods and gave it to humanity, endowing our species with mastery of one of Nature's most awe-inspiring powers, a mastery that placed us above all other living beings. But as it is often said and equally as often forgotten, with power comes responsibility. The power to control fire meant that humans had to choose how to use it: either to create or to destroy. Zeus, who disliked any threat to his dominance, chained Prometheus to a rock where an eagle devoured his liver each day, which then grew back again each night. Prometheus's agony ended only when Hercules came to his rescue. This myth is an early exploration of the conflict between religion and science, pitching reason against faith: the more humans know about Nature and its resources, the less room there is for belief in the supernatural. The control of fire makes humans less like animals and more like gods, a very dangerous status for immature creatures with a primitive capacity for moral judgment. The power to control Nature teaches us nothing about how or whether this power should be used. The moral dilemma of how to use scientific knowledge is as present today as it was then—with vastly more urgent consequences.

The pre-Socratics attempted to undermine the power of myth with a new tool—dialectics, the art of investigating the truth of an argument through reasoned discussion. By choosing rational debate over dogmatic faith-based belief, these early Western philosophers planted the seeds of what would become science two thousand years later. They shifted the cultural focus of their time from stories of gods and their deeds to the workings and mechanisms of the natural world. They also suspected that often things are not what they seem. Uncovering the secrets of Nature and its inner workings became their quest, driven by an obsession to find the truth about the world. Through the fog of magic and divination prevalent in those days, the pre-Socratics sought the power of an alternative way of knowing anchored on the natural and knowable as opposed to the supernatural and unknowable.

To understand how innovative these thinkers were, step back into the past and try to visualize the Cosmos as people did twenty-five centuries ago, without what we know now. Their main observational tool was the naked eye. They had no telescopes or detectors, only very rudimentary tools such as the gnomon—a rod lodged vertically in the ground, used to tell time by the position and length of its shadow (used, for example, in sundials).

Looking at the night sky during a moonless night, they saw countless dots of light, just as we do when far away from artificial lights. They noticed that some celestial lights blinked while others did not. Curious, they wondered what those lights were and why they disappeared during the day. They realized that the whole night sky rotates from east to west, just like the Sun does during the day. With patience, they noticed that some celestial lights, the ones that didn't blink, moved slowly across the sky with respect to the blinking lights. They were called *planetes*, from the Greek word for wanderer. Planets, they reckoned, were wandering celestial lights. The other lights, the ones that seemed fixed with respect to each other, were *asters*, or stars. Some stars seemed grouped into patterns identified with images of animals, of magical creatures, of gods, of geometric figures—what we call constellations. These "fixed stars" moved together as a whole, their scintillations oblivious to time like little light diamonds encrusted on the dark celestial dome. This whole majestic structure of stars and planets rotated about the Earth.

The centrality of the Earth, the ground these early thinkers stood on, seemed obvious and inevitable. And doesn't it still if we forget what we now know? We see the skies rotating about us, not us about the skies. We don't feel dizzy as we do on a merry-go-round. It is thus not surprising that early maps of the Cosmos placed the Earth at the center of everything. Earth was special. It was different from the celestial luminaries above. It didn't glow on its own. As early as 450 BCE the Greek philosopher Empedocles proposed that Earth and everything on it were composed of four basic elements—earth, water, air, and fire—mixed with each other in different proportions. Very reasonably, the world was made of the kinds of stuff that we can see and touch, although, predictably, philosophers before and after Empedocles disagreed on the specifics, as they did about the stuff that made the stars and planets.

There was also the problem of time. The lights in the heavens didn't seem to ever change. Down here, however, everything seemed to be in a constant state of flux. The elements mixed to create all sorts of nonliving and living concoctions: wet dirt, dry sand, dusty wind, clouds and fog, glowing hot coal and metals, trees, insects, birds, snakes, horses, people. <u>The apparent timelessness of the celestial luminaries clashed with the ever-changing nature of things on Earth.</u> Down here, nothing was eternal; up there, everything appeared to be. Was time, then, relegated only to the earthly realm? Were the heavens timeless? Taken together, the list of properties that made Earth different, even exceptional, was getting longer—not just its central position in the Cosmos and its material composition, but also the fact that time and change appeared to be particular only to the terrestrial reality. Earth, these early philosophers understood, was the realm of the mortal, of aging and decay. But also of birth and regrowth, of chance and the unexpected. Despite all the challenges that time's passage brings, at least it grants the privilege of witnessing a rose bloom or a rainbow weave the sky with color, even if only for a brief moment.

And then there was the problem of being human, which, of course, is still very much with us. We are a strange kind of animal, endowed with a capacity for complex symbol manipulation and an urge to make sense of the world. Why are we aware of the incessant torrent of emotions and thoughts that floods our minds? Our ancestors drew on cave walls and built weapons and totems, wondering about (and fearing) the workings of Nature with a deep sense of awe and reverence. Fast-forward thousands of years, and we still draw, and build, and ponder. Why are we so different from other animals? To what purpose?

To answer these questions, our ancestors told stories of Creation, mythic narratives that, as mentioned above, played many roles, including setting us apart from the rest of the natural world. Creation stories are usually about us, about how we came to be. Their specific environmental framing mirrored the realities of the storytellers. Desert dwellers told stories of life fashioned from dirt and mud; cultures surrounded by ocean revered water and Sun; if from cold weather, the stories were of ice and fire; if from jungles, of trees and rain. There was the world of the seen, the reality perceived by the senses, and the world of the unseen, the mysterious forces that seemed to drive much of what happened with powers beyond the conceivable. This polarization between the seen and the unseen emerged from an understanding of reality based solely on our senses, an understanding that fractured the world into two conflicting realms: the seen and known, and the unseen and unknowable. In this dual framework, there was yet no room for the unknown, that which could in principle be understood through a process of inquiry and analysis. To our ancestors, our powers rested only within the realm of the known, the natural world that we could act upon, the human sensorium. Still, even within this limited realm of the known, we could do much, imposing our will over that of other creatures through the use of fire, tools, and skilled strategy.

This power, encompassing what we could control of the world, eventually led—especially in Western cultures—to a deep-seated belief of human and Earth exceptionalism: we humans are at the apex of Creation; Earth is at the center of the Cosmos. This belief magnified when the great monotheistic traditions conflated the centrality of the Earth and the superiority of humans as part of God's plan. Our supremacy was inevitable. Worse, it was sacred.

The belief that we are at the apex of Nature became deeply ingrained in our cultural norms. For most people, it is the dominant worldview, and thus very difficult to change. But changed it must be. We need a new narrative of who we are and where we fit in the natural world, a narrative that is less about domination and more about belonging. In this, we have much to learn from the pre-Socratic philosophers. As we will see, they might have disagreed about the details of their world systems and the nature of change, but many, departing radically from previous mythic narratives, suggested a deep connection between humans and all forms of life as either emanating from a single primal substance or jointly emerging from the mixing of a few. More important, some even proposed that this ongoing process of creation from and degeneration into primal materials included not only what was on Earth, but also what was in the skies. Recognizing that there were similarities between what happened in the skies and down here, these thinkers suggested that the whole Cosmos danced to the tune of change and transformation, following natural, as opposed to supernatural, rules. What had up to then been a rigid duality of the known and the unknowable when describing the natural world cracked open to give rise to a third possibility: the unknown, encompassing that which could be probed and understood through rational discourse, expanding the realm of the possible. Between the immediacy of the senses and the unapproachability of the divine was a physical world, waiting to be deciphered.

An important aspect of early pre-Socratic philosophy aligns it with Indigenous cultures from around the globe. To many Greek thinkers Nature was alive, an organism pulsating with the energy of living matter. Thales of Miletus, who flourished around 650 BCE and is considered the first of the pre-Socratics, is credited with saying that "all things are full of gods." The gods here are not the humanlike gods of Mount Olympus, like Zeus and Hermes. They are the forces hidden within living and nonliving objects, responsible for their physical properties, such as, for example, the powers of the magnesian stone, able to attract metal (what we call magnets today).

Aristotle, writing more than three hundred years after Thales, attributed to him and his followers a sort of animistic take on Nature, the belief in a soul-like spirit that pervades all things. To Thales and his followers, Nature was alive, in a constant state of flux and transformation. Matter changed shape and properties but kept its inner essence. For comparison, consider the modern way we describe water changing into ice or into steam, all made of the same H_2O molecule, but with different properties due to changes in temperature and pressure. The "gods" in all things was an attempt to make sense of the hidden forces responsible for such material changes.

The first pre-Socratics believed in a unifying principle for all matter, that all that existed derived from this one source and reverted to it in due course. What that primordial element was varied from thinker to thinker. For Thales, it was water. For his disciple Anaximander, an abstract substance he called the

Apeiron, the Greek word meaning indefinite or boundless. For his follower Anaximenes, it was air. For all of them, matter danced its choreography of creation and destruction, but now without gods in control. Natural processes unfolded without divine conductors. In a profound change of worldview, the forces that animated things remained mysterious but were no longer part of a supernatural mythology. The unknowable became the unknown, opening the Universe to scrutiny and analysis, heralding the dawn of scientific thinking in the West. Echoes of the remarkable pre-Socratic imagination resonate with many of the questions we still ask today.

THE FIRST COSMOLOGIST

Anaximander's first ever mechanical model of the Cosmos was a huge first step toward the transition from mythic to rational explanations of natural processes. Back around 600 BCE, his goal was to describe the world of experience by means of concrete mechanisms. He is (probably inaccurately) credited with the invention of the gnomon and of being an expert at building sundials and celestial globes. Anaximander is also credited with being the first to draw a map outlining land and sea. Although none of this is confirmed, there is no question that he believed in the power of tools and models to describe reality as perceived by human eyes. His pragmatism was grounded on his belief in the Apeiron, the abstract primal substance that gives rise to all that exists, including the heavens and the worlds in them.

Anaximander devised a cosmology of an eternal Cosmos with worlds coming into being and perishing in endless succession "according to necessity; for they pay penalty and retribution to each other for their injustice according to the assessment of Time." This quote is the only confirmed surviving fragment of his writings, a poetic evocation of the finiteness but also the connectedness of all existence, from celestial worlds to living creatures, under the cold assessment of time's passage. The Apeiron "enfolds all and steers all," as Aristotle later wrote, being the connective fabric of all existence, living and nonliving.[1]

A profound and inspiring beauty emerges from Anaximander's vision. All that exists, living and nonliving, originates from the same primal substance, the Apeiron. When Apeiron becomes a material entity, its existence is limited by the passage of time. In due course, it will perish and recycle its materials into other forms of being. A material object, be it a world, a rock, a plant, a wave, or a person, is an ephemeral identity in the perpetual ebbing and flowing of the primal substance.

In Anaximander's vision, the stars were no longer mysterious lights in the celestial dome. He made the stellar patterns into a mechanism. There were wheels spinning around the Earth. These wheels had rims filled with fire. The Sun, the Moon, and the stars were fire spewing from holes in the wheels. As the wheels turned, the holes turned with them—what we on Earth see as the celestial objects spinning around us.

The Cosmos became a machine built of wheels within wheels. What before was a mystery now became a mechanism, a rational cosmology replacing a mythic narrative.

Some Greek and Roman authors of antiquity, including Plutarch (c. 46–c. 120 CE), attributed to Anaximander a cosmogonical model, a description of the origin of worlds. What is striking is how Anaximander intuited some aspects of planetary formation in ways that resonate, at least in a figurative way, with our modern understanding of how planets, moons, and stars are born:

> [Anaximander] says that which is productive from the eternal hot and cold was separated off at the coming-to-be of this world, and that a kind of sphere of flame from this was formed round the air surrounding the earth, like bark round a tree. When this was broken off and shut off in certain circles, the Sun and the Moon and the stars were formed. (K&R 131)

There was a primal formless material (the Apeiron) that combined all opposites—chaos. Then chaos self-organized into order, without the action of a god. The separation of hot and cold stuff from this "productive" mass ignited the process of creation, generating a sphere of flame that surrounded the Earth like "bark from a tree," reminding us of rings of matter that form during planetary formation. These rings of fire composed Anaximander's mechanical model that depicted the Sun, Moon, and stars as fire coming out of holes in wheels surrounding the Earth.

Worlds form from a primordial material that separates and reorders itself into fiery rings that eventually become the celestial luminaries. It is uncanny that twenty-five centuries ago someone imagined such a richly dynamical mechanism for the formation of worlds. What matters is not how accurate Anaximander was compared with modern science, but how some of the key steps he envisioned remain valid today, including the notion that the Sun, planets, and moons emerge from the same primeval fireball of stuff.

Modern astrophysics tells us that stars are giant gravitational engines that transform hydrogen—the most abundant chemical element in the Universe—into all other chemical elements, from the chemical components of minerals that make rocks to the calcium in our bones and the iron in our blood. Stars are born when contracting hydrogen clouds become dense enough to ignite the fusion of hydrogen into helium at their core. They live dramatic lives and die dramatic deaths, spilling their entrails through interstellar space, seeding stellar nurseries with their chemical remains in a dance of death and resurrection. Again and again through cosmic history stars cycle hydrogen into the stunning variety of atoms and molecular compounds that make up all that exists. Had Anaximander known of this, would he have named hydrogen his Apeiron?

Before we move on from Anaximander, it's worth pointing out that he seemed to have dreamed up the precursor of what we nowadays call the *multiverse*, the hypothetical collection of possible universes that includes our own. The specific details are nebulous, and there is much discussion among classicists about Anaximander's take on the creation and destruction of worlds. Still, experts agree that in his cosmology worlds emerge from and perish back to the Apeiron in the "assessment of Time." The controversy is in the details of what these "worlds" are. Some say that Anaximander meant the cycle of continuous creation and destruction of many worlds coexisting in space, but others say he meant that the cycles apply just to our world being created and destroyed in time.

These two interpretations resemble, in some qualitatively suggestive ways at least, the kinds of multiverse that are being proposed (more on them soon). Briefly, a multiverse is a collection of universes, each, at least in theory, with different physical properties. In one, the electron has a certain mass; in another, a different mass. Or, from universe to universe, the strength of gravity and other forces may vary.

According to current physics models, a multiverse can exist in space or in time. A multiverse in space describes a collection of universes that coexist in space, like soap bubbles floating about, even though they don't communicate with each other. You can't travel from one universe to another without violating a few laws of physics. Some universes may live a long time; others will perish after a short time. Our Universe (capitalized, to differentiate it from others) is one of these coexisting universes, the one that happens to have the physical properties that allow for it to be old enough and filled with the right kinds of matter to form galaxies made of billions of stars and planets going around most of these stars. Among this vast plurality of worlds, at least one supports a spectacularly abundant biosphere that includes one terrestrial animal species capable of complex language and endowed with a deep urge to understand its origins.

A multiverse in time describes a single Universe that goes through (possibly) endless cycles of creation and destruction, like the mythic phoenix. In some models, the physical properties of the Universe can change from cycle to cycle.[2] We happen to exist because in the current cycle the properties allow for the formation of stars and planets, and for life to emerge and evolve.

Fewer than two centuries after Anaximander, these first notions of worlds coming into being and perishing greatly expanded with Empedocles, with the Atomist philosophers Leucippus and Democritus, and most explicitly, with Epicurus. Some of their intuitions were visionary.

LOVE AND STRIFE

As we follow the thinking of the pre-Socratics, we see a growing interest in trying to develop natural mechanisms that would describe not just the nature and arrangement of the celestial objects, but also how things are organized down here on Earth, including, starting with Empedocles (c. 494–c. 434 BCE), the origin of living creatures. Again, the remarkable innovation here is the absence of divine intervention as the operative principle behind the emergence of the natural order, for both nonliving and living matter.

Empedocles is generally considered the first to invoke the coexistence of the four basic material elements that we attribute to Greek thinking: earth, water, air, and fire. Whereas his predecessors proposed a sort of unified theory of matter—a "monism" based on a single material substance as the fundamental one (water, air, fire, the Apeiron)—Empedocles suggested that all four elements coexisted in the cosmic sphere, attracted and repelled to one another by imbalances in the amount of Love and Strife: too much attractive Love, and Strife would come in to separate the materials; too much Strife, and "a soft, immortal stream of blameless Love kept running in" to join things together again (K&R 331). Existence results from a delicate balance between the two.

To Empedocles, Love and Strife, attraction and repulsion, are the forces that organize matter in the shapes we see. Even living creatures emerge from disjointed parts of bodies that are brought together or separated by the interplay of the two interactions: "Many things were born with faces and breasts on both sides, man-faced ox-progeny, while others again sprang forth as ox-headed offspring of man, creatures compounded partly of male, partly of the nature of female, and fitted with shadowy parts" (K&R 337). What a nightmarish scene this is!—bits of animals coming together with bits of humans to form failed creatures. But we identify here the early seeds of evolutionary thinking, portraying life as an experiment in survival of the fittest—the ones with the proper forms and proper balance between Love and Strife. As Aristotle wrote later, summarizing Empedocles's thinking, "Wherever, then, everything turned out as it would have if it were happening for a purpose there the creatures survive, being accidentally compounded in a suitable way" (K&R 337). Note how "accidentally compounded in a suitable way" brings together the notions that the forms of living creatures are randomly compounded and that there is a "suitable" way for them to assemble, that is, the one that leads to successful survival. Things were no longer "full of gods" but responded to tendencies to attract and repel as matter organized in different shapes. This was true for worlds and animals alike, a unifying mechanism acting throughout the Cosmos that transformed the primal materials into things and things into primal materials. A few decades after Empedocles, the rise of Atomistic ideas radicalized the rupture with divine intervention.

ATOMS, THE VOID, AND MANY WORLDS

The view that Nature's essence was change and transformation was not left unchallenged. Many pre-Socratic thinkers vehemently disagreed, suggesting instead that truth was to be found in that which didn't change, the eternal Being. This is the age-old rift of Being versus Becoming: Where do Nature's secrets hide? In the myriad transformations we witness, or in some deeply hidden and unchangeable reality? Parmenides, who lived about one hundred years after Thales and Anaximander, would say that if we are interested in what *is*, we shouldn't be focusing on what changes; after all, if something changes, it becomes what it is not and hence can't be fundamental. He goes on to suggest that what *is* cannot come into being. Why not? Because the process of becoming something is change and hence transformation that leads to another identity. So, the fundamental stuff of reality is timeless and can't be subdivided. It must be everywhere, fill everything, and just be. Parmenides reasoned that the Cosmos must be spherical in shape, the most perfect of proportions, devoid of stress on any point. He went on to say that the changes we perceive with our senses are deceitful illusions that cloud our judgment. After all, what kind of final truth can we attach to a reality that can be altered by drinking a few glasses of wine or ingesting hallucinogenic plants?

The choice between Being or Becoming was the challenge philosophers faced around 450 BCE. The brilliant solution, as is often the case, saw beyond this false dichotomy. Better than choosing between the two is to combine them. This is where the first Atomists come in. Aristotle credited Leucippus with being the originator of the idea that all that exists consists of indivisible little bits of stuff called atoms. Then, his pupil Democritus took the idea and ran. The atoms themselves cannot change—they are little chunks of Being—but they can combine to form the shapes we see in Nature. So, we have Being turning into Becoming in a cosmic game of Legos. Granted, such atoms are far from Parmenides's perfect unchangeable sphere of Being that permeates all of reality. But how else could one make sense of the two conflicting notions, given the transformations we constantly witness around us? The Atomists focused more on making sense of reality as they could see it and less on metaphysical questions. As Democritus apparently said, "In reality we know nothing: for truth is in the depths." The mystery of existence hides deep in the fabric of reality.

If there were early suggestions of a plurality of worlds in the writings of Anaximander, Anaxagoras, and Anaximenes, the Atomists cemented it. According to the classicists G. S. Kirk and J. E. Raven, "they are the first to whom we can with absolute certainty attribute the odd concept of innumerable worlds (as opposed to successive states of a continuing organism)" (K&R 412). Leucippus and Democritus claimed that reality consisted of atoms moving in a void of infinite extension (the "whole"). The atoms themselves were also infinite in number and in kind. Playing with the idea of infinity opens up all sorts of possibilities. If there are infinite atoms and they make worlds by forming swirling vortices where they collide and join each other, then it follows that there should be an infinite number of worlds. Each world is isolated from the outside by "a circular 'cloak' or 'membrane' which was formed by the hooked atoms becoming entangled" (Aetius, quoted in K&R 410).

The atoms of the Greeks bear little resemblance to modern atoms, but the notion that matter is made of little building blocks is still very much present in physics. Although modern atoms are divisible (made of "up" and "down" quarks and electrons) and not infinite in number (there are ninety-two naturally occurring atoms and a few more made artificially in laboratories), the idea that matter is composed of elementary particles—tiny indivisible material bits—remains essential, being the driving concept of high-energy particle physics. And although the modern mechanism of stellar and planetary formation differs from atoms swirling in vortices, we do know that the coalescence of matter through a combination of rotation and gravitational attraction forms rotating proto-planetary disks that become stars and their orbiting planets. The intuition of the Atomists is nothing short of spectacular.

About a century after Democritus, Epicurus (341–270 BCE) took the Atomistic perspective to new heights, suggesting that the creation and destruction of worlds was a natural consequence of a Cosmos devoid of divine intervention. According to Epicurus, the gods may have existed, but they were utterly indifferent to human affairs or the running of the cosmic machinery. As Mary-Jane Rubenstein wrote in her excellent compendium of multiverse ideas, "given infinite time, infinite substance, and infinite space, any material configuration that can emerge, will."[3] This includes our world, even if it may look so special and finely tuned for life.

In a Cosmos of infinite possibility anything that can happen, will happen.

However, considering infinity led Epicurus to an important conclusion, elaborated in his *Letter to Herodotus*: "There is an infinite number of worlds, some like this world, others unlike it."[4] This is not the first time Earth appears in the infinity of time, an idea that apparently echoes Democritus. But contrary to Democritus, Epicurus realized that even if the number of atoms is infinite, for *this* world of ours to reappear, there should be only a *finite number* of types of atoms. Otherwise, if the number of types of atoms was infinite, why should any world appear more than once? The sheer number of combinations and possibilities would make the probability of two identical worlds reappearing effectively zero.

We identify here the germ of what will later crystalize as the Copernican revolution, the notion that Earth is not special in any way. Not only was it assembled accidentally by the coalescence of atoms, but as it turns out, it will reappear on and on in the infinitude of time, together with an infinite number of other worlds, some similar to ours and others quite different. The early pre-Socratic exceptionalism of our world vanishes in the dust of Earth's infinite reappearance in time and multiplicity across space.

Still, a key difference between Epicurus and Copernicus is that for Copernicus the Cosmos was finite and spherical. His revolutionary idea was to displace Earth from the center of Creation, a nonissue for Epicurus, given that an infinite space can't have a center. For Copernicus, Earth was just another planet revolving around the Sun. The notion of other Earthlike worlds was far from his mind. For Epicurus, ours was one of many Earths spread across the vastness of space. Copernicus probably would have balked at this idea, given that he believed there were only six planets in the whole Cosmos, all of them revolving around the Sun. For both, however, Earth had lost its special status. We must now reverse this mindset and restore the essential status of Earth in the Universe. Only when we retell the story of our planet will we be able to change how we relate to it.

LIBERATION COSMOLOGY

The pre-Socratic Cosmos was godless. That was its most essential rupture with any other Greek school. Plato's cosmology had a Demiurge, a cosmic crafter who molded matter into the shapes we see in the skies, who framed order from chaos. The Demiurge wasn't an all-powerful creator; it used material that already existed to fashion worlds. Aristotle's divinity had a different function. His "Unmoved Mover" was the First Cause, the originator of the first push that set the Cosmos in motion. Being the First Cause meant that it couldn't have been caused by anything before; it had to be uncaused, a sort of deity that just existed in time as pure knowledge, who had the power to turn his timeless Being into the motion of everything else that existed within the Cosmos. To Aristotle, the Cosmos was a huge machine of spheres within spheres—a cosmic onion, the planets and other celestial bodies attached to a few of them, all rotating in different ways to reproduce the motions we see in the skies. He imagined this Unmoved Mover to act from the outside in, the initial push coming from what later became known as the Primum Mobile, the first motion sphere, the outer boundary of the physical Cosmos that Dante placed as the ninth heaven in his *Divine Comedy*. Outside of it remained only the Empyrean, the tenth sphere, the realm of God and the elect.

The Atomists would have none of this. The notion of a divine creator implies hierarchy and submission to a supernatural power. To the Atomists, this is where superstitious fear and enslavement to ritual originate, belittling humanity and limiting our freedom. Furthermore, against direct evidence, it implies that the gods actually care about human affairs, whereas in reality they seem to be utterly indifferent, given the amount of strife and suffering in our world. In addition, they argued, if a god created our world as an abode for humanity, why should so many regions be inhospitable to us? Surely there wouldn't be so many deserts and frozen landscapes. As the Roman poet Lucretius (c. 94–c. 55 BCE) argued in *The Nature of Things*, the epic poem that gave renewed voice to the Atomistic worldview, if the world was created for humanity, Nature wouldn't be in a permanent state of war against us:

> *What's the reason Nature multiplies and feeds*
> *The enemies of man on land and sea—the bristly breeds*
> *Of wild brutes? How does it come about Disease abounds*
> *At the change of seasons? Why does Death make his untimely*
>   *rounds?*[5]

Only cruel gods could have done such a thing. And why would they? To have fun at our expense? To watch our unjust suffering as a kind of blood sport? The Atomists claimed that belief in a supernatural power hinders our freedom to become whole and embrace responsibility for our lives. To be free meant to be free of god fear. To them, there were no gods of any kind and, if there were, they were inaccessible to us and indifferent to our plight.

In 1629, Rembrandt painted a self-portrait, *The Young Rembrandt as Democritus the Laughing Philosopher*, a rare case where the usually somber Rembrandt depicted himself smiling. Democritus was known as the Laughing Philosopher, free as he was from belief in the supernatural and attuned to what truly mattered: the mechanisms of the world and the search for human self-knowledge.

Nothing that existed in Nature was eternal, be it here or anywhere in the Cosmos. Lucretius's *The Nature of Things* is an eloquent embodiment of the Atomists' worldview, prescient and inspiring, a major driving engine of the Renaissance, as Stephen Greenblatt proposed in his 2011 Pulitzer Prize–winning book *The Swerve: How the World Became Modern*.[6] Embracing Anaximander's vision of beginnings and endings "according to the assessment of Time," Lucretius goes on to argue that as sure as the ongoing dance of life and death recycles the matter that makes our surroundings, distant worlds in the heavens are continually being born and destroyed:

> *Those elements we see comprise the Sum of Things—since they*
> *Are made of substance that is born and that must pass away,*
> *We must conclude the nature of the whole world is the same.*
> ....
> *Thus when I see even major members of the world consumed*
> *And born anew, then earth and heavens, it must be assumed,*
> *Also have a birthday, and in time to come are doomed.*[7]

According to this worldview, gods do not interfere with the process of the creation and destruction of material things on Earth or in the heavens. The cycles of aggregation and dispersion of matter follow natural rhythms, atoms coming together to form new objects and worlds, only to spread apart and be recycled into new objects and worlds in the eternity of time. The Earth too had a birthday and, "in time to come," will be "doomed." In the eternal dance of existence, the atoms that make our world, as well as everything existing in it—rocks, frogs, butterflies, clouds, you—

will, in the fullness of time, be embedded in other worlds and in things in other worlds.

We carry the stuff of the Cosmos in us. We are the stuff of the Cosmos. And while we are alive—and only while we are alive—we know this. We can see why Democritus and Lucretius made Rembrandt smile.

THE STOICS AND THE MULTIVERSE

The Atomist Epicurus had an enemy, the Stoic philosopher Zeno of Citium. Even if they agreed that only through cultivating an understanding of Nature one could attain a state of "calm imperturbability and the living of the simple life,"[8] they vehemently disagreed about pretty much everything else, from the constitution of matter to the existence of other worlds. If, for Epicurus, matter could be divided all the way to small indivisible atoms, for Zeno it was a continuum, meaning it could be divided to no end. Space, too, was not a void where atoms moved but, echoing Aristotle, was filled with a primal substance, a sort of fiery ether that changed intensity depending on how dense it was. This primal fire was the tool that a divine intelligence, a god-architect, used to fashion the world. If, for Epicurus, there was an infinity of worlds emerging and perishing by the constant coming together and dissolution of atoms, for Zeno Earth was the only world in the Cosmos, a world with repeating cycles of existence. Zeno believed that our world would end when the Sun consumes the Cosmos in flames, a death by fire. But then, phoenix-like, from the ashes of a world consumed a new Cosmos would emerge, together with a new Earth, only to perish again in due time, in an eternal cycle of creation and destruction.

This process, called *ekpyrosis* ("out of fire"), is reminiscent not only of Empedocles and his cycles of creation and destruction from the tension between Love and Strife, but venturing farther east, of the rhythmic creation and destruction of the Universe through Shiva's cosmogonic choreography. In contrast with the Atomistic coexistence of many worlds being born and destroyed in infinite space—a multiverse in space—it introduces another idea: the creation and destruction of the *same* universe in infinite time—a multiverse in time.

Both types of multiverse reappeared in late twentieth-century cosmology, dressed in the mathematical language of Einstein's general theory of relativity, from efforts to bring the four known forces of Nature into a unified framework, the so-called unified field theory. We now consider models of multiverses in space, derivative of a string theory landscape, and multiverses in time, derivative of models called bounce cosmological models. There is a lot to unpack here, so let's go by parts.

WHAT'S A FIELD?

Fields are the lifeblood of fundamental physics. In philosophical parlance, fields are the ontological substrate of reality, the stuff that makes up everything. To get a feel for a field, consider a fridge magnet. As you bring it closer to the fridge's door, the magnet is pulled toward the door—even without touching. The closer the magnet is to the fridge, the stronger the attraction between the two. Why is that? A field is a manifestation in space of a physical disturbance. Just like fire is the source of the heat we feel around it, every field has a source. In the case of the fridge magnet, the magnetic material generates a "magnetic field" that extends to the space surrounding it. A piece of iron placed near it will respond to the magnetic field by being attracted to or repelled by it. A field is like a ghostly presence originating from a source and spreading to the space around it. Its intensity drops with the distance, the precise ratio depending on the kind of field and the type of source. The Sun attracts the planets, and the planets attract the Sun through their gravitational fields. Any object with mass in the Universe attracts everything else through gravity, as Isaac Newton proposed in 1687. Gravity acts everywhere. It shapes solar systems and galaxies. It connects the whole Cosmos, controlling the expansion of the Universe. Newtonian gravity unified terrestrial and celestial physics, showing they obey the same laws. If our bodies can't reach for the stars or for the atoms, our minds can. This is, perhaps, the most wonderful aspect of science—to bring the unreachable closer to us.

You are attracted downward by the Earth. And you are also attracted by the Moon, the Sun, this book and all objects in your surroundings, the Andromeda galaxy, and a quasar billions of light-years away. And you attract them back. "When we try to pick out anything by itself, we find it hitched to everything else in the Universe," wrote the naturalist John Muir in *My First Summer in the Sierra*.[9] Gravity embodies deep interconnectedness between all things, the invisible arms that embrace you, our world, the Universe.

Einstein was an heir to Pythagoras and Plato. He believed that geometry was the language of Nature and that Nature was fundamentally rational. He firmly believed that the human mind could, through rare bursts of creativity, glimpse some of this underlying order. Newton's theory of gravity was efficient but mysterious, with gravity acting at a distance across empty space. How could the Sun dictate the orbits of planets from so far away? What was it about mass that created such attraction? Newton wouldn't speculate. "I do not feign hypothesis," he famously wrote. An alchemist and a devout Christian, he deferred the source of this attraction to powers beyond the material: "This most elegant system of the Sun, planets, and comets could not have arisen without the design and dominion of an intelligent and powerful being."[10] To Newton, the nature of gravity was inextricably entwined with God's presence in the Cosmos.

Einstein picked up where Newton left off, proposing a very rational (and beautiful), albeit mysterious, explanation: an object with mass creates a gravitational field that bends space around it. The mass is the source of this field. The more massive the object, the more distorted the space around it. In Einstein's theory, what Newton called gravitational force was simply motion on a bent space. Like a child going down a slide, planetary orbits are the most energy-efficient paths around the Sun. (These paths are called geodesics.) Gravitational attraction results from the local curvature of space around an object with mass. It is a manifestation of the gravitational field.

WHAT'S A UNIFIED FIELD THEORY?

Aside from gravity, we currently know of three other forces we call fundamental. Electromagnetism is the more familiar one, being a joint manifestation of electricity and magnetism in motion. We experience electricity in many ways, for example when we witness a lightning strike or get shocked touching a metal doorknob on a dry winter day. And we are also familiar with magnets. We know that a magnet creates a magnetic field around it—like the fridge magnet. But if you now move this magnet, something wonderful happens: a moving magnet creates an electric field around it. In turn, electric fields make electric charges move, creating electric currents. This is why hydroelectric dams use spinning magnets to generate electricity. Likewise, in a spectacular feat of complementarity, an electric charge creates an electric field around it. And a moving electric charge (accelerating on a straight line or along a curved path, or perhaps oscillating) creates a magnetic field. Motion unifies electricity and magnetism into electromagnetism. Then magic happens. As the electric charge oscillates—think of a bobbing cork in a tub of water—its changing electric and magnetic fields, its electromagnetic field, propagate outward as waves, like the water waves from the bobbing cork. What's the magic? This waving electromagnetic field spreads out with the speed of light. It *is* light. We call it electromagnetic radiation because light is only the tiny portion of electromagnetic radiation human eyes can see—visible electromagnetic radiation. Light is an oscillating electromagnetic field propagating in space. In empty space, light travels at 186,000 miles per second. You blink your eyes and light goes seven and a half times around the Earth. As far as we know, nothing in the Universe can move faster. Wherever there is light, there are electric charges dancing. As light bounces around in space, reflected, refracted, diffracted, we capture some of this dance with our eyes. And what our eyes can't see, our instruments can. The world is illuminated. Our narratives of reality are the tales this dancing light tells across space and time, reflecting off the face of someone we love, refracted by a dewdrop on a rose petal, furiously shining from burning stars.

The last two fundamental forces are less conspicuous, even if equally essential—the strong and weak nuclear forces. They act at the nuclear and subnuclear realms, deep within the atoms that make up matter. The strong force is the silent sculptor of the depths, responsible for keeping atomic nuclei together and for keeping quarks within protons and neutrons. Without it, there would be no atoms, no matter, no us. The weak force is the purveyor of change within the elementary particle realm, responsible for radioactivity and radioactive decay, often present when some kind of transmutation happens deep within the atomic nucleus. The weak force morphs protons into neutrons (more accurately, up quarks into down quarks). And that's only one of its claims to fame. Stars like our Sun are giant nuclear fusion engines, the cosmic alchemists that churn hydrogen into helium for billions of years. As the furious process consumes the star's entrails, the weak force sees that energy is released—the energy that warms our faces, drives the weather, and powers our solar panels. By orchestration of the weak force, stellar nuclear fusion releases trillions upon trillions of ghostly particles known as neutrinos, capable of crossing whole planets without stopping. Right now, as you read this, trillions of neutrinos originating at the core of the Sun are going through your body—every second. An invisible bridge of neutrinos links us to the heart of the Sun. "What is essential is invisible to the eye," said the fox to the Little Prince, in Antoine de Saint-Exupéry's fable. As with love and friendship, there are layers of reality that escape us, invisible to the eye but equally essential to our existence.

The project of unification relies on the belief that these four forces—these four fields—are manifestations of a single force, the unified field. We look at reality with myopic eyes, and it is this shortsightedness that precludes us from seeing the unity of the four fields in all of its grandeur. If only we could look into the depths, we would see the world anew, unified and whole, reality expressed as a mathematical masterpiece encoded in geometric language.

It sounds like a prayer, a Plato-inspired devotional to a geometer god. Earlier in my career I was a devout follower of this worldview. But with time I experimented with research in diverse topics in theoretical physics, and my worldview gradually changed. Although I recognize the aesthetical appeal of such a worldview and its honorable intellectual lineage from Pythagoras to Einstein, I also recognize its fundamental inconsistency with the way science actually works. I summarized my position more than a decade ago in a book critiquing the belief in a final unification of the forces of Nature—*A Tear at the Edge of Creation*[11]—and offering arguments that are not central to our concerns here. What is central is the conceptual impossibility of realizing such a quest and, more important, its connection to a worldview that amplifies the Copernican principle to cosmological scales, in the guise of the modern concept of the multiverse. Let us address each of these two points in turn.

WHY THE CONCEPT OF FINAL UNIFICATION IS INCONSISTENT WITH SCIENTIFIC METHODOLOGY

There are many reasons for the impossibility of a final unification. An obvious one that is usually ignored is that we cannot be sure that there are only four fundamental forces of Nature. What if a new instrument reveals evidence for a fifth? Or a sixth? They would have to be included in this unified theory. The point is that statements about the finality of knowledge are wrong, more the product of hubris than logical consistency. What we see of the world relies on the instruments we use to amplify the parts of reality we can reach. It follows that if we can't see all, we can't know all. And we can't see all. Omniscience is the business of gods, not humans. Instead, we should celebrate our remarkable achievements without trying to transform science into the oracle of final truths about reality.

Science is a human construction, a self-correcting, ongoing narrative of the world. There is no end to this search, given that new discoveries usually come with new, previously unknown, sets of questions; I described it in a metaphor in *The Island of Knowledge*:[12] If what we know fits on an island, the island of knowledge grows as we learn more. But the island is surrounded by the ocean of the unknown, the realm of the not-yet-discovered. As the island grows, so does its periphery, the boundary between the known and the unknown. Knowing generates not-knowing. Discoveries generate new questions. As long as we continue to explore and ask questions, the ocean of the unknown will grow. Ergo, there can't be a final unified theory, even if its scope is limited to the interactions of elementary particles of matter.

The best we can hope for is to build (temporarily) successful models of what we know of physical reality. Hints of unification should be considered as partial results, given what we don't and can't know.[13]

THE GROWTH OF COPERNICANISM AND THE LIMITS OF SCIENTIFIC KNOWLEDGE

The previous paragraphs set the mood for what follows, since most multiverse models of today are derived from attempts to unify the four forces of Nature. Contrary to their Greek antecedents, they are firmly grounded in modern scientific reasoning. They are fully hypothetical, but the hypotheses that lead to the notion of a multiverse are based on extrapolations from two currently tested theories: the Standard Model of particle physics, which describes impressively well our experimental knowledge of how elementary particles of matter interact up to the very high energies achieved at the giant particle collider at CERN, the European laboratory of high-energy physics near Geneva, Switzerland; and the Standard Model of cosmology, which describes incredibly well how our expanding Universe emerged from a hot and dense primordial soup of elementary particles 13.8 billion years ago and became populated with galaxies and stars.

Note the words "standard" and "model" in both. The two theories are our current "standards" of what we know of particle physics and cosmology. They are open to modification, and we physicists look forward to modifications, as they point toward new scientific discoveries. They are also "models," meaning incomplete descriptions of physical reality, simplifications we build to encode what we know. Scientific models are maps of the natural world, *not* the world itself. To substitute maps for reality is not only conceptually wrong, but potentially dangerous, as I have explored with my colleagues Adam Frank and Evan Thompson in *The Blind Spot: Why Science Cannot Ignore Human Experience*,[14] and will get back to later in the context of alien life and the search for Earthlike planets. The philosopher of science Edmund Husserl called the identification of the map with the territory "surreptitious substitution," the confusing of a model of the world with the world itself. For example, the fields we described above represent how we humans model what we can observe and measure of the effects of attraction and repulsion between objects. The question of whether fields "exist" or not brings us back to the nature of physical reality and is a complicated one. Many physicists would say, "Of course fields exist. We measure them with our instruments!" But fields are not the measurements; they are the interpretation we give to the data we are able to gather with our machines. This is a very important distinction.

We measure phenomena in Nature and build models to describe what we measure. Occasionally, we need to change our models to include new experimental results and observations. New science is built from the failure of old science. But when new experimental results take too long, theoretical physicists take the lead and extrapolate current models even before results contradict them. They know that current models are incomplete (always!) and want to extend them to embrace new physics. This is how it should be, as long as the extrapolations are acknowledged as such, that is, as long as physicists don't lose sight of where their hypotheses are coming from—not from certainty, but from uncertain and untested extrapolations. Extrapolations are maps of unknown territories. And we know that maps of unknown territories must be used with much care, or we get lost. Unfortunately, that is often not the case, and speculative ideas gain the status of being so compelling as to be inevitably correct. This is problematic because it circumvents the usual scientific methodology of empirical validation and gives the public the illusion that we know much more than we actually do. Speculate we must, but with care and humility. Scientists shouldn't sell maps of the world telling people how to go from point A to point B if they don't know whether point B even exists.

The current success of our two Standard Models (particles and Cosmos), coupled to their limitations, has justly inspired a huge extrapolation industry into what happens at extremely high energies near the Big Bang, the event that marked the beginning of time. We do want to know how particles of matter interact at extremely high energies and the properties of the Universe close to the Big Bang, but we don't have the exploratory tools to probe these realms. What are we to do?

This is the dilemma of extrapolation: we can fill the gaps in our current knowledge only with our current knowledge. The seventeenth-century French philosopher Bernard Le Bovier de Fontenelle considered the conflict between curiosity and shortsightedness to be the essence of philosophy (or of any pursuit of knowledge, science included): we know only what we know, but we want to know more than we do. So, we plow ahead, pushing forward into uncharted territory.

In 1543, Copernicus posited that the Sun and not the Earth

was the center of the Cosmos (see chapter 1). He had suggested as much in an earlier document from 1510 that he circulated among his peers. This rearrangement of the solar system demoted Earth from its presumed centrality, correctly turning it into another planet orbiting the Sun. After Kepler, Galileo, Descartes, and Newton cemented this idea as the proper architecture of the solar system, the accessible Cosmos grew in size, as increasingly larger telescopes allowed for a clearer view of the skies. More planets were discovered (Uranus in 1781 and Neptune in 1846), as were more moons circling these worlds. Copernicanism became the notion that Earth is just another world, not central or essential in any way to the cosmic machinery. It could exist or not, as Jupiter could exist or not. Earth and, by extension, its denizens were not special in any way. The mechanistic science of the eighteenth and nineteenth centuries depicted our planet as a tiny speck in the vastness of space, irrelevant in the big scheme of things. This worldview inspired the Enlightenment, given that this eighteenth-century intellectual movement predicated reason above all else as the path to individual freedom and society's progress away from monarchy and political and religious dogmatism. Since human reason determined Earth's place in the Cosmos and the laws of Nature with such success, it gained credibility in all spheres of human endeavor, becoming the moral compass guiding humanity's progress.

Gradually, and in particular following the Enlightenment, a broader value judgment was attached to our unimportant position in the skies, and Copernicanism became more than just a rearrangement of the solar system. It became a statement about our planet's mediocrity. Voltaire's brilliant satirical tale *Micromégas*, from the middle of the eighteenth century, encapsulates this view well, exploring, among other themes, the smallness and triviality of Earth and of humanity against two extraterrestrial giants, one from Saturn and the other from a planet orbiting the star Sirius (Micromégas).[15] This worldview cared not for the spectacular confluence of physical, biochemical, and geological properties that conspired to generate Earth's stunning biosphere. It cared not for the natural world and the life in it, except as an object of rational study. It pushed aside with indifference and arrogance Indigenous cultures and their wisdom, casting their deep connection to the land as primitive and uncivilized. As Western culture distanced itself from a spiritual connection to the planet, the machinery of industrialization took the world to be humanity's possession, a thing we could exploit with impunity for our own material benefit. Detached from Nature, mechanistic market forces set forth to fuel progress from Earth's entrails—the oil, the gas, the coal—without ever pausing to consider the consequences of this rampant destruction of the natural environment. The cities, the global transportation system, the goods we own—the world of modernity was erected from the degraded remains of life buried for millions of years under our feet.

During the past centuries, Copernicanism expanded from an astronomical fact about our solar system into a worldview whereby the more we learn about the Universe, the less important we become. The Sun is just an ordinary star located some twenty-seven thousand light-years away from the center of our galaxy, the Milky Way. The Milky Way itself has between one hundred billion and four hundred billion stars, most of them with orbiting planets. Although we can't pinpoint a precise number, we can estimate with some level of confidence that, between planets and their moons, there should be more than one trillion different worlds in our galaxy alone.

Let's pause for a second to consider the enormity of this. One trillion worlds, each different, each with its own history, its own chemical composition, its own geophysical and orbital properties. *Comparative planetology* is an emerging methodology that helps us make sense of this vast diversity of worlds, grouping them into different categories. Which worlds are rocky like Earth and Mars, and which are gaseous like Jupiter and Neptune? What are their masses and chemical compositions? What are their sizes and at what distances do they orbit their parent star? Do they have mountains, lakes, oceans? Could they host living creatures of some kind? And if so, how would we know?

We will address these questions in detail, but for now, we explore the growth of Copernicanism from our solar system outward to larger and larger cosmic distances. In 1924, the American astronomer Edwin Hubble showed that the Milky Way is but one among billions of other galaxies in the Universe. In 1929, he showed that these galaxies are moving away from one another, a discovery we now call the expansion of the Universe. As is often the case in the history of science, these remarkable findings owe much to a powerful instrument, the one-hundred-inch reflector telescope atop Mount Wilson, outside of Los Angeles. Endowed with plasticity, space stretches and carries galaxies along, like corks floating down a river. This cosmic drift may continue forever, as galaxies move farther and farther apart while their stars exhaust their fuel and gradually fade away; or it may reverse itself, turning cosmic expansion into cosmic contraction. Behind the equations describing the comic dynamics, we can hear echoes of the ancient Hindu myth of the dancing god Shiva, who creates and destroys the Cosmos in endless cycles.

We can't know for sure the faraway fate of the Universe. Shockingly to many, and despite numerous statements to the contrary, the fate of the Universe is unknowable. At this point, it is worth briefly digressing on this topic, focusing on what it tells us about the power and limitations of the scientific enterprise.

To predict with confidence the far distant future of the Universe we would need to know two things we can't know. First, we would need to know the extreme long-term properties of all that exists in the Universe. Currently, we believe that there are two main contributors that fill up the void of space, apart from the ordinary matter that makes us and stars, rocks, the clouds of Jupiter, and the more exotic elementary particles catalogued in the Standard Model of high-energy physics. Called dark matter and dark energy, their nature and material composition remain unknown, despite decades of intense experimental search and theoretical work. The "dark" in their names designates their otherworldly property of not emitting any kind of visible light or invisible electromagnetic radiation. We know that dark matter and dark energy exist because their gravitational forces act on the shiny stuff that we can see. Dark matter acts on galaxies and clusters of galaxies, and dark energy acts on the Universe as a whole, affecting the way it expands. Mind-bogglingly, familiar ordinary matter contributes only 5 percent of what's out there, with dark matter and dark energy filling up the other 95 percent.

We expect to learn much more about them in the coming decades, including what dark matter is (it may be more than one thing, such as swarms of exotic particles or, instead, particles gathered into lumpy balls—my favorite since I contributed to the theoretical discovery of one of the current candidates, called oscillons) and whether it is stable, that is, whether it survives through the eons of time without decaying into something else.[16] We equally expect to learn more about dark energy within a similar time frame, as powerful new telescopes begin to map how it evolved along billions of years of cosmic history, measuring whether it has changed in time or remained essentially unchanged through the eons.

I hope to witness such progress. Still, even if we discover the nature of dark matter and dark energy, to make a very long-time prediction about their behavior is extremely difficult since it requires very good statistics of what we are observing (swarms or lumpy balls, and their average lifetime) and the ability to observe their behavior for arbitrarily long times. In simplistic terms, to know that something will live forever, you need to live forever too. Statistical inference is a very powerful tool indeed, but it's, well, statistical. We can't ever be sure which particular history will unfold, only what the possible histories would be. Probabilities tell many stories—some more ordinary and others rarer. We can't know for sure how the story of our Universe will end. What we can do is extrapolate from current knowledge without any guarantee of absolute certainty.[17] This attitude requires what is known in philosophy as *epistemic humility*, that is, to accept the limitations of what we know and can know. Epistemic humility should not be confused with epistemic nihilism, which says we know nothing, which is just silly.

The second limitation to our ability to predict the long-term fate of the Universe is related to instrumentation, the tools we use to make measurements and that determine what we can conclude from these measurements. As German physicist Werner Heisenberg of uncertainty principle fame once wrote, "What we observe is not Nature itself but Nature exposed to our method of questioning."[18]

Every piece of information we gather from the natural world is filtered through our senses. We experience reality before we can measure it, and this experience depends in fundamental ways on the sensorial machinery belonging to the human animal. What we can directly sense of reality is but a small sliver of what's out there. Invisible presences are all around us. And I'm not talking about ghosts or spirits, but myriad kinds of electromagnetic waves we can't see, sounds we can't hear, objects too small or too far away for our senses to capture, like the cosmic rays and solar neutrinos that fly through our bodies. Our instruments are *reality amplifiers*, tools that detect, magnify, and translate natural phenomena in ways that we can capture with our senses. We don't see electrons or ultraviolet radiation, but we do see gauges and color screens and blinking lights, we hear ticks and tocks and buzzes, and we smell and touch and taste.

Technological innovation goes hand in hand with this amplification of physical reality. Just think that what Galileo could see with his telescope—an instrument that changed our worldview forever—today we can easily see with a pair of good-quality household binoculars. Our particle detectors can sense collisions between elementary particles of matter, allowing us to dive into the reality of subnuclear physics. Our giant telescopes allow us to observe nascent galaxies billions of light-years away, while gravitational wave detectors allow us to capture the minute vibrations of the very fabric of space from black holes colliding at vast astronomical distances.

We build our maps of reality from the fragments of the world we are able to capture with our instruments. As our instruments improve, so do our maps.

But as the Argentinian writer Jorge Luis Borges cautioned in his masterful one-paragraph short story "On Exactitude in Science," no map can be a perfect representation of the territory unless it is as big and detailed as the territory itself.[19] And of what use would a map like that be for us? The power of science is not in representing Nature as it is—which I'm arguing is an impossibility anyway—but in describing Nature as we experience it. We mustn't forget that every measuring instrument or detector has a precision limit, a range, a finite amount of resolution. We see only the world that our instruments allow us to see. It then follows that even if we extrapolate forward to a future of great technological innovation and inventiveness, there will always be aspects of the world that will escape us. The far distant future of the Universe is beyond our grasp, a humbling and unavoidable consequence of how science works. Only our imagination can go there.

Beyond the long-term fate of the Universe lies the multiverse, the ultimate expression of Copernicanism. With the multiverse, not even our Universe is special, being just one among a multitude of others. But before we get into a nihilistic funk of cosmic proportions, we must investigate the modern ideas of the multiverse and to what extent we should consider them a serious threat to our cosmic standing.

THE MULTIVERSE AS THE GOD OF THE GAPS FOR PHYSICS

We have seen that the idea of the multiverse is not new, having been present in the thoughts of Greek philosophers who lived more than two thousand years ago. Atomists and Stoics argued whether a multitude of worlds continually emerged and collapsed in the vast expanse of space or whether ours is the only world, undergoing its own creation and destruction in the vast expanse of time. As we have noted, both scenarios have returned in the framework of modern cosmology. Although there are different ways to motivate the notion of the multiverse in modern cosmology, they all depend on extrapolations from current physics to a realm very close to the Big Bang and thus extremely remote from what we can test experimentally.

In string theories, the multiverse emerges as a form of landscape of possible universes, with a lowercase u as these universes are not our Universe. According to string theory, space has more than the three dimensions we observe, and these extra dimensions can bend, fold, and have complicated topologies, while also being extremely tiny and thus inaccessible to any of our instruments that measure physics at subnuclear scales. String theories build a compelling connection between the geometry of these extra dimensions and the values of the constants of Nature, numbers we measure in our three-dimensional Universe, such as the speed of light, the mass of the electron or of the Higgs boson, or how strongly these and other particles interact with each other. Through a process called spontaneous compactification each particular shape of the extra dimensions generates a set of constants of Nature with specific values in our three-dimensional reality: geometry would thus predict the kinds of possible universes, ours being one of them. Ideally, and this is why I was so taken by the theory during my early career, out of the many possible shapes of this extra space, the theory would *predict* the one that corresponds to our Universe. In other words, if the theory were correct, we could *deduce* from geometry the physical properties of our Universe, the ultimate Platonic dream, the ecstatic realization of a two-thousand-year dream of unveiling Nature's deepest structure through human reason, the climactic meeting with God's mind. The string landscape is the abstract space formed by all these possible geometries and related compactifications, with dips here and there corresponding to a different universe. These universes exist outside each other; traveling between them is forbidden by the laws of physics. Thus, if you exist inside one universe, you cannot directly reach or detect other universes. It is as if we existed inside a fishbowl, incapable of knowing what lies outside.

The string landscape is as compelling as it is problematic. Physics is first and foremost an empirical science, relying on testable hypotheses that are amenable to validation or refutation. Therefore, an untestable, unobservable multiverse lies outside the realm of ordinary science. A multiverse is very different from the idea of an atom, which remained untested for millennia until it was confirmed, or of the Higgs boson, which was discovered over five decades of remarkable efforts. Atoms and particles belong to our physical reality and can be detected. If the multiverse lies beyond what is observable not in practice but in principle, how does it belong to physics and not to another kind of dialectic argumentation?

Ideas have been proposed to indirectly test the existence of other universes. For example, if a neighboring universe collided with ours in the past, the collision could have left a specific imprint on the microwave background of photons that permeates the sky.[20] Such patterns have been searched for and not found. And even if they had been found, their observation would at most constitute very indirect evidence that other universes existed. How could we be sure that no other physical processes would have been capable of producing similar patterns? Any hypothesis we propose now depends on our current knowledge of physics being extrapolated to realms well beyond what we can trust. The existence of other universes falls within the dictum that Carl Sagan made popular in his TV series *Cosmos*—"Extraordinary claims require extraordinary evidence"—but that has roots going back as far as at least the early 1700s.[21]

Some scientists go as far as to propose the multiverse as an alternative to God. The reasoning is that the multiverse offers an explanation as to why the constants of Nature conspire to create the Universe that we exist in, capable of having planets that harbor living creatures. Given the absurd number of possible universes within the string landscape, our Universe was bound to be one of them—the winner of the cosmic lottery, at least from the perspective of beings that play the lottery. This sort of argument is a revival of the (re)tired "God of the gaps" theological argument, which placed God in the gaps of our ignorance about the workings of the Cosmos. Newton, for example, attributed the stability of planetary orbits to divine interference. The multiverse wants to reject the idea that the constants of Nature are finely tuned to produce the Universe we exist in. Everything seems to work as if by magic, resulting in our being here. So, was it random chance or purposeful architecture? In the absence of a scientific explanation for the values of the fundamental constants in our Universe, which, by the way, was the original motivation for string theory, there must have been a fine tuner, that is, a divine architect. For this reason, the argument goes, science's best way out of this conundrum is the multiverse: our Universe is the product of chance, one dip in the vast string landscape. No need for the Architect. However, to go from the lack of a scientific explanation for the measured values of the constants of Nature to the existence of a divine tuner as the only alternative is completely unjustified. Who decreed that science must explain the numerical values of the constants of Nature, that it can successfully address this kind of question? Witness the God of the gaps argument, but now curiously used backward by some scientists to justify a scientific hypothesis that can't be empirically validated, that is, the existence of the multiverse. One colleague famously quipped that "if you don't want God, you'd better have a multiverse." This is not science but theology, and bad theology at that. It is a false dichotomy. The multiverse plays the same role as a God of the gaps, a scientific argument that, since it's untestable, can't be ruled out by science, just like God can't.

The alternative is to consider the constants of Nature as the physical parameters we measure and use to build our narrative of the world. They are the alphabet of physics, the scaffolding supporting our mathematical maps of reality. They are not constants "of Nature" but of our human mapping of physical reality as we sense and measure it. They don't belong to the Universe. They belong to us. The maps we make, wonderful and compelling as they are, are not the territory. What's truly magnificent about science is not that it can allow us to know everything—a premise that doesn't make sense anyway—but that it allows us to see so much.

Echoing the Greek Atomists, the string landscape is a multiverse in space. Other cosmological models echo the Stoics and their *ekpyrosis*, proposing a multiverse in time: there is one Universe that undergoes an endless succession of cycles where matter and energy are squeezed to huge densities (the beginnings) followed by expansions and then by new contractions. Known as *bounce* or *cyclic models*, they have the clear advantage of not depending on the many assumptions needed for string theories to be viable, such as extra dimensions of space. They do, however, evoke highly speculative physical processes that currently lack observational evidence.[22]

MEDIOCRITY AND THE NEED FOR A POST-COPERNICAN REVOLUTION

We now circle back to Copernicanism and its implied indifference toward our existence, one of our central concerns. The Copernican principle has spawned another principle, the *mediocrity principle*, which extends the irrelevance of our cosmic position to the commonality of life and even intelligent life in the Universe. According to the mediocrity principle, the mediocrity of life on Earth stems from the fact that the laws of physics and chemistry are the same across the Universe. It then *assumes* that, by extension, so are the laws of biology based on Darwinian evolution by natural selection. If our planet is nothing special, and life emerged here from nonlife some four billion years ago and evolved to become intelligent, advocates of the mediocrity principle claim that not just life but also intelligent life would have emerged on countless other Earthlike worlds across the Universe. It thus follows that Earth is mediocre, or just ordinary, and life is mediocre, intelligent life is mediocre, we are mediocre.

This kind of extrapolation is as misleading as it is dangerous. It is misleading because it is scientifically incorrect. It is dangerous because it leads to a neglect of our planet and trivializes life, intelligence, and the essential role our species plays in cosmic history. The Copernican principle rests on an undeniably correct astronomical fact: Earth is a planet orbiting the Sun like other planets in our solar system. Anything else that follows from this fact to inform the mediocrity principle is the result of a philosophical viewpoint based on little more than scientific hubris, not on well-grounded science. To wit, the mediocrity principle rests on three fundamental assumptions: (1) there are many Earthlike planets in the Universe, meaning planets capable of spawning and sustaining life for the long haul; (2) life emerges on many of these planets; and (3) on a considerable number of them, life evolves to become intelligent.

Of these three assumptions, the only one that currently has some level of observational support is the first one, although it also weakens under close scrutiny. As we will see in detail in part II, there have been many observations of planets orbiting other stars, known as *exoplanets*, that indeed seem to be rocky and within the habitable zone of their host star. (The habitable zone of a star delimits the range of orbits where liquid water is possible on the surface of a planet, as we will see in chapter 4).

However, a planet being rocky and a planet being Earthlike are very different things, given the many complex geophysical properties a planet needs to support life for long enough periods of time. As an obvious example, Mercury, Venus, and Mars are rocky planets but certainly not Earthlike in their abilities to spawn and sustain life. In fact, the meaning of "Earthlike" is currently not very precise, focusing mostly on physical properties of the planet, such as having a mass and radius similar to those of Earth and orbiting within the habitable zone of its host star. But since life is an essential part of being Earthlike, this current qualitative denomination is lacking in specificity. Mass, radius, and orbiting within the habitable zone of its host star are the minimal requirements for a planet to be Earthlike, but surely they are not sufficient. A truly Earthlike exoplanet would also need to have an atmospheric composition very similar to Earth's, a composition that indicates an active biosphere—a much more stringent requirement.[23]

As for assumption numbers 2 and 3, we have little to no understanding of how life emerged from nonlife on Earth and how it evolved to become intelligent, given the many contingencies of our planet's unique history. The origin of life remains a mystery, while the evolution of simple unicellular life to intelligent life is not a necessary—and certainly not an inevitable—pathway for the evolution of life. Life cares about whether it is well-adapted to its environment so that it can reproduce successfully, not whether it can build rocket ships or write poetry. In other words, although intelligence is clearly an evolutionary benefit, there is no guarantee that life will go there. Dinosaurs existed for more than 150 million years, undergoing many mutations. However, they didn't become intelligent, at least not the kind of intelligence that builds technological civilizations. Current science doesn't justify extrapolating from the existence of other rocky planets in our galaxy and beyond to having many instances of intelligent life in the Universe. Life, and much less intelligent life, does not simply follow from favorable astronomical conditions. This trivialization (or mediocritizing) of life and intelligent life has serious consequences for how we see ourselves and the planet we inhabit and share with other life-forms. The way we tell the story of who we are matters.

The relevance of our existence is not merely a scientific question related to our cosmic location. It is an existential question, a question that relates to our values and moral standing and that asks for multiple perspectives, science being but one of them. It follows that we shouldn't decide whether or not we matter in the grand scheme of things on the basis of a faulty scientific argument that objectifies our home planet and deems our species as a mere link along a causal chain of events. Unfortunately, our current cultural narrative, which tells the story of who we are, has bundled the two together, and the cosmic irrelevance of our location (which, it turns out, may also be disputable) has been inflated to signify the irrelevance of our planet, of the existence of life here, and of our species and other complex life-forms.

This narrative must be overturned, just as Copernicus and his supporters overturned the centrality of our cosmic location more than three centuries ago. As with any story, we have a choice of how to tell it. Following Copernicus and the Enlightenment, we have told our story as one of increasing smallness: the more we know about the Universe, the less important we become.

To advance toward a value-free universality, science distanced itself from spiritual concerns and focused on a quantitative, data-driven description of the natural world. This is, of course, how science operates. Spirituality in science lives within the subjective worldview of scientists who consider themselves spiritual, not in the science they produce. It may inspire and inform scientists' worldviews, but not their scientific output. Einstein, for example, believed in a Spinozan rational divine presence in Nature, but you won't find mention of this in any of his scientific papers.

But our story is not just a scientific narrative. It embraces multiple cultural dimensions, science being one of them. There is no denying that the expanding Universe is vast and that our planet is but a little speck in an ordinary spiral galaxy. But this is only part of the story. For on this speck, life emerged and evolved to spawn a species capable of self-awareness, with a spiritual thirst to know its origins, its destiny, and the meaning of being alive. This, in itself, is grounds for a profound sense of awe that goes well beyond the facts of cosmic distances and sizes, or even of how many Earthlike planets are out there. Our scientific accomplishments shouldn't make us lose sight of who we are. Quite the opposite, it is by looking with a different mindset at what we have learned so far that our existence, and that of rare planet Earth, gains a new level of relevance. We must retell our story within this new perspective to rescue our cosmic identity. We can go somewhere new as a species only if we rethink who we are. And given the current state of our project of civilization, we must urgently rethink who we are. In a post-Copernican worldview, science embraces meaning to reorient our collective future. To begin, we must look at the Universe with different eyes.

PART II

WORLDS DISCOVERED

CHAPTER THREE

THE DESACRALIZATION OF NATURE

> The eternal silence of these infinite
> spaces fills me with dread.
>
> —Blaise Pascal, *Pensées*

THE FIRST TRANSITION: HOW EARTH LOST ITS ENCHANTMENT

Before we could see other worlds through telescopes or land on the Moon and send spacecraft probes to the confines of the solar system and beyond, we reached out to the vastness of space with our imaginations. As a child, I used to spend the summers at my grandparents' home in the mountains about two hours outside Rio de Janeiro, where I was born and grew up. Surrounded by gardens and fruit trees, their house was my magical place, a refuge from the crowded, noisy, and polluted (but still stunningly beautiful) big city. Although Nature was controlled and tamed, its presence was powerful in my grandparents' yard, bursting with flowers and insects and myriad birds. Like instruments in a symphonic orchestra sounding melodic phrases that together become a moving work of art, the tropics reverberate with an explosive life energy embodied in the countless living creatures populating the ground, the waters, and the air. A thriving tropical ecosystem is a giant experiment of the biologically possible realized through plants, animals, and fungi. I didn't know it then, as I collected and catalogued beetles, spiders, and butterflies and identified birds, bats, and frogs, but that small piece of the planet was my shrine, a portal to the sacredness of the natural world.

Years later, after my grandparents passed away, the house was sold. The first thing the new owners did was to cut down the pines and magnolia trees to clear the land. That desecration left a hole in my heart that is still open today, half a century later.

The air was cleaner then, and the moonless nights darker, with little influence from bright artificial lights. On hot summer nights I'd lie down in the grass with my cousins in awe of what was above. We were children of the first Moon landing, which we watched with minds blown and eyes glued to my uncle's black-and-white TV as Neil Armstrong took the first human steps on another world. For billions of years, life on this planet had been tethered to its atmosphere. But now, in an extraordinary transition, life had unleashed itself toward outer space, in search of its cosmic origins. Vast as it is, the Universe became a little smaller. If science and the human imagination could take us to our satellite world 240,000 miles away, it could go anywhere. We would watch the sky in silence, scanning for shooting stars, wondering whether other kinds of creatures were looking at us from faraway worlds, asking, as we did, if they were alone in the Universe. I'm still asking that same question.

Being social animals, humans have a hard time tolerating

loneliness. We need to belong to a group to find a sense of purpose. This remains true as our community circles grow from family and friends to schools, clubs, and churches, and from there, to states and countries. The last link in this chain of belonging, still missing from our collective awareness, is to expand our community circle to embrace the whole natural world. To live a full life, we must find a place and a sense of purpose at every layer of these concentric circles of community, even if we may invest more energy in specific layers. But purpose cannot benefit only the individual. A selfish sense of purpose brings no community fruit. It brings loneliness. Purpose may spring from the self, but if it remains centered on the individual, it will bring isolation and not belonging. A bright light surrounded by mirrors on all sides does not illuminate the outside. When we expand this thought from the inner community of family to a global community that embraces the whole natural world, when we see humankind as a single species inhabiting a small planet, we can attain a new sense of purpose, working together as planetary citizens. The alternative, as we can clearly see around us, is planetary neglect. If we neglect our families, we will end up with no family. If we neglect our friends, we will end up with no friends. If we neglect our communities, we will end up with no community. And if we neglect our planet, we will end up with no planet. At least no planet we can inhabit.

We evolved to crave the comforting security of a group, of belonging to a community that gives us a sense of identity and purpose. We need groups that recognize us as valued members. One powerful aspect of all religions is to offer companionship through a shared set of values. This is also true for secular communities bound by a sense of allegiance. In groups we find strength, we feel protected, we find purpose in relating to and helping others. Our inner light spreads to our surroundings. The challenge of our times is to find ways to expand this sense of purpose and belonging to the planetary community, to embrace the life collective.

The path forward is not easy. But we can find guidance from two sources that, at first sight, seem extremely different from each other. We can find guidance from those who preceded the machine-driven technological civilizations of the West—the Indigenous cultures that, for millennia, have worshipped Earth as an enchanted realm. And, perhaps surprisingly to many, we can also find guidance from our current scientific narrative, bringing together what we have learned from the physics of the very small, the atoms and molecules that we and everything else are made of, to the very large, the astronomy of stars, exoplanets, and the Universe as a whole. However, for the current scientific narrative to guide us toward a planetary community, we must reframe its focus—not by portraying science as the triumph of human reason over the natural world, but as a narrative that situates humanity within the epic vastness of cosmic and evolutionary history.

※

Our sense of community had its origins in the bands of hunter-gatherers that preceded agrarian civilization. Quite possibly, we inherited such characteristics from previous hominin species, from Australopithecus to Neanderthals. What we do know that differentiates *Homo sapiens* from previous species is a largely amplified capacity for symbolic imagination and representation due to our enlarged frontal cortex. We attach symbolic value to objects and represent them through art and language. This capacity allows our mapping of the world to become at once concrete and abstract, as we identify forces we can and can't control as being responsible for animating existence. To our ancestors, reality was at once that which they could act upon and control—hunting, picking fruit from trees, making fires and having babies, moving elsewhere in search of water and protection—and that which was beyond their reach, the mysterious forces behind the phenomenal world—the day and night cycle, unpredictable volcanic eruptions and mighty storms, the strange urge of living things to survive and spread across the land.

To have some level of control over such mysterious forces, our predecessors built a bridge between the concrete and the abstract that connected the real and the magical aspects of the world. Nature became divinized, filled with spirits that infused all that existed. Our animistic ancestors didn't distinguish between the natural and the supernatural. There was no boundary between the concrete and the magical. Spirits were as real as trees and mountains and waterfalls, being inseparable from them. Indigenous traditions across the globe relate to forests, valleys, and mountains as relatives, as uncles and aunts, and to the land as a mother. The dead are as present as the living, invisible to the eye but not to the heart. Members of these communities respect the natural world as they would a member of their family. The bond between humans and Nature is the root of Indigenous cultural identity and moral landscape. Plants and animals are as entitled to the land as humans. People share with animals the need to forage and to hunt, not being above or below them. Indigenous cultures see themselves as belonging, together with all life, to the sacred land. The land doesn't belong to people; people belong to the land. This moral hierarchy—Nature before people—is essential to how Indigenous cultures relate to the natural environment around them. The respect and unique intimacy that comes from a reverential kinship with the land is profoundly different from the objectification of Nature that followed the growth of agrarian societies across the globe.

The Brazilian Indigenous leader and activist Ailton Krenak puts it clearly:

> *For a long time, we were fed the story that we, humanity, stand apart from the great big organism of Earth, and we began to think of ourselves as one thing, and Earth, another: Humankind versus Earth ... My community with what we call Nature is an experience long scoffed at by city folk. Rather than see any value in it, they poke fun at it: "He talks to trees, he's a treehugger; he talks to rivers, he contemplates the mountains." But that's my experience of life. I don't see anything out there that is not Nature. Everything is Nature. The cosmos is Nature. Everything I can think of is Nature.*[1]

You are never alone when the world is your family.

An agrarian society takes possession of a piece of land to provide sustenance to its population. The land becomes human property, "my piece of the world." The moral hierarchy is reversed—people before Nature.[2] If the land is fertile and the planting successful, the community grows, as does the need to create rules of behavior that protect the social order. There is also a need to enforce such rules in case of disagreement. Small or large, any community of humans needs such legal structures. As the number of inhabitants around agrarian centers grew, the need for enforcing the rules also grew.[3] Authority had to be absolute, top down, and not a matter of public discussion. With the need for authority came the need to justify it. The solution was to separate the people from the land, severing the ancestral kinship between humans and Nature. The spirits of the forest, of the waterfalls, of the mountains, of the skies, had no place within the walls of growing cities. City walls kept the undesirable out, including intruders from the natural world, be they real-life predators or magical creatures. Within the city and its surroundings, the ruler needed authority beyond dispute, a divinely granted top-down authority. The more powerful the gods, the more powerful the ruler who represented them on the ground: "My king is more powerful than your king because my god is more powerful than yours."

The advent of organized religions—both polytheistic and monotheistic—created a sharp division between our world, the natural world, and the world of the gods, elevated to the "supernatural," or the beyond-the-natural realm. The laws down here didn't apply to the otherworldly realm of the gods. Time and space, the limits of our longevity and the hardships of our motions through the land, didn't affect them. The gods existed above Nature. Rulers of the people were their emissaries, their power justified by divine authority. As the Roman emperor Constantine the Great or, as late as the early 1700s, the French king Louis XIV, the Sun King, believed, the gods sanctioned their power. Some cultures went even further. In Egypt, the pharaoh was considered to be a god on Earth. Even when there were no such claims, alliances between state and church forged a divinely justified power hierarchy at the expense of a growing estrangement between humans and the natural world.

As the gods departed the realm of the living—the valleys, the forests, the mountains—successful agrarian societies became fast-growing cities, an agglomeration of humans that pushed Nature outside its boundaries. Satellite farms tamed natural resources to serve the needs of the people, while cities became trading posts for buying and selling goods and crafts. The "wilderness," the wild expanses of the world, was tagged as dangerous, unpredictable, home to beasts of prey—and was to be avoided or, when needed, exploited and destroyed. Gradually, the age-old sacred alliance between humans and Nature turned into an open conflict. Nature, once the life-giving mother goddess, became Nature, the enemy. Ancestral religions that worshipped Nature were deemed pagan and sinful. Worse, with the expansionist colonization of Western powers, native religions were deemed "primitive" and their followers "savages." To "civilize" someone meant forced conversion to Christianity or death. The "great conversation," to use Catholic priest and ecotheologian Thomas Berry's inspiring expression, was no longer between humans and Nature but between humans and an absent God.[4]

With the rise of monotheistic faiths, redemption and solace became abstract pursuits, removed from a spiritual connection with the natural world. The land had lost its enchantment. God was far removed from Earth, inhabiting the unreachable confines of Heaven. Prayer and ritual now aimed to connect with the ethereal realm of God and the elect, while the land, the forest and its mysteries, became associated with darkness and decay, home to the unruly, to the temptations of the flesh, to the seductive magic of evil spirits.

The sense of communion with the divine, once horizontal and concrete, anchored to a spiritual connection to the natural world to which we and all living creatures belong, turned vertical and abstract, anchored instead to the supernatural notion of heavens above, the realm of God, detached from the human condition, from the hardships of a flesh-and-blood existence in a changing world. By the fifth century, Saint Augustine and other Christian theologians completed the transition. How could the harshness of the world compete with the promise of eternal heavenly bliss? Even the few outcasts who worshipped Earth as part of God's creation—the desert fathers and mothers, ascetic sages and mystic saints—immersed themselves in Nature seeking for communion with an abstract God. Miracles became ruptures with the physically possible, temporary divine interventions on Earth from a supernatural distance.[5] The gods were gone from the land, and Earth lost its magic.

THE SECOND TRANSITION: FROM A CLOSED COSMOS TO AN OPEN UNIVERSE

The Church's early embrace of the Aristotelian model of an Earth-centered, onionlike Cosmos made a lot of sense. After all, Aristotle proposed a clear division between what happened on Earth—the terrestrial realm—and what happened to the Moon and above—the celestial realm. The central Earth was the seat of change and transformation, with all matter composed of combinations of the four basic elements: earth, water, air, and fire. The Moon and all other celestial luminaries were eternal and unchanging, composed of the fifth element, ether. Eight spheres circled the central Earth: one for the Moon, one for the Sun, five for the visible planets, and one carrying the stars. Beyond the sphere of the stars, the loving and good Unmoved Mover originated all cosmic motions from the ninth sphere, called Primum Mobile ("first motion"). Finally, the last sphere, the Empyrean, was the realm of God and the elect, where Dante Alighieri placed his Paradiso. The Christian medieval Cosmos was closed, static, and spherical.

This rigid cosmic structure began to crumble after Copernicus proposed his heliocentric model. At first, few listened to his ideas, as we have seen. But about one hundred years after the publication of *On the Revolutions of the Heavenly Spheres*, Kepler, Galileo, Descartes, and Newton completed the transition from a geocentric to a heliocentric Cosmos, leaving Aristotle's ideas behind while reinventing physics in the process.

As might have been expected, the Earth's removal from the center of Creation caused great confusion for theology and for science. Contrary to what many think, however, Copernicus didn't suffer any harsh censorship from the Church.[6] This tragic distinction belongs to the one-time Dominican friar Giordano Bruno, who had actively defended heliocentrism since the 1580s, echoing Epicurus in suggesting that stars were other suns with planets rotating about them, many of them crowded with life of their own. Bruno was an embattled visionary who dreamed of uniting the warring Christian faiths by sprinkling elements of Hermetic mysticism into the boiling pot of the Reformation and Counter-Reformation. The results were disastrous. Beyond his defense of Copernicus, the Church took offense with Bruno's contrarian views on eternal damnation, the Trinity, the divinity of Christ, and the virginity of Mary. After eight years of trial, the Inquisition condemned an unrepentant Bruno as a heretic. He was burned at the stake at Rome's Campo de' Fiori square in 1600. Visitors enjoying the busy market at the square can't avoid the somber statue of a cloaked Bruno, today celebrated as a martyr for intellectual freedom. At the statue's base we find the inscription, "To Bruno—The Century He Divined—Here Where the Stake Burned."

Given this terrifying precedent, a vastly more cautious and politically savvy Galileo decided to openly repent his defense of heliocentrism in 1633, avoiding torture and the stake. Instead, he was sentenced to daily prayer and house arrest for the remaining days of his life. One of his two daughters, both nuns, performed most of the prayers on his behalf. Not to be silenced, Galileo's disciples dodged the Church's censorship and smuggled his books and ideas across Europe. Thanks to Galileo, physics became the mathematical study of motion and its laws, while his pioneering telescopic evidence dealt a devastating blow to Aristotle's Earth-centered Cosmos.

Meanwhile, in Prague, Kepler established the three mathematical laws of planetary motion, combining data and theory to build the conceptual foundations for a heliocentric Cosmos. Earth and the other distant worlds surrounding the Sun obeyed the same laws of orbital motion. Taken together, Galileo and Kepler showed that motions on Earth and in the heavens obeyed precise quantitative laws, laws that reflected an order that extended across the whole Cosmos. What for the Greeks had been the realm of philosophical argumentation became the study of physical reality, amenable to quantitative description.

This remarkable intellectual achievement gradually erased the sacred value our ancestors attributed to our home planet, as the nascent science of the heavens legitimized Earth as a rocky planet orbiting the Sun. From an astronomical standpoint, there was nothing exceptional about it. Religion amplified this view, given that our presence here made Earth even less attractive, with our propensity for materialistic decadence and the pleasures of the flesh. An ordinary rocky planet inhabited by sinners couldn't hold much magic.

The first decades of the seventeenth century were transitional times, when the new role for Earth was counterbalanced by the age-old belief in a finite Universe enclosed within the Empyrean. God was out there on the tenth sphere, distant but present. His role was confusing. Catholics believed he could act by a miracle in the world, while Protestants believed in the miracles of the New Testament and Jews in the miracles of the Hebrew Bible. For all monotheistic religions, God's removal from the world required faith in God's abstract presence. Such was the prevailing belief in European cultures at the time. By contrast, as we have seen, many Indigenous cultures held Nature itself as divine and considered being in the world a privilege deserving of devotion and reverence. There was no separation between the living world and the spirit world. Faith is mostly needed for religions in which gods are removed from the realm of the living.

Then, something remarkable happened: Isaac Newton. Born in 1642, the year Galileo died, Newton quickly became the leading figure in the newly emerging physical sciences. But contrary to Galileo and Kepler, he saw the physics of the heavens and the physics of terrestrial phenomena as one and the same. Gravity, in particular, was the great unifier, the force responsible for the motions of falling objects on Earth and for the planetary orbits around the Sun, Earth included.[7] It was also responsible for the tides, for the recurring orbits of some comets (Halley included), for the tilted spinning of our planet (the precession of the equinoxes), and for its slightly oblate shape. Gravity was the universal sculptor, acting on grains of sand attracting one another at a beach, extending its reach all the way to the magnificence of the planetary motions and beyond, to the stars. Gravity enveloped the whole Cosmos with its embrace. What many people don't know and schools don't teach is that, to Newton, gravity was inseparable from God.

Being a theist, Newton believed God to be immanent in the world, his omnipresence ensuring the stability of the whole of Creation. Every object with mass attracted every other object with mass. You attract the Andromeda galaxy, and the Andromeda galaxy attracts you. Through gravity everything is connected, even if these connections weaken with (the square of the) distance, quickly becoming negligible. Andromeda and you are linked, but so very feebly as to not cause a noticeable effect. But the arms of gravity do extend outward and reach out to everything in the Cosmos, shaping the Universe itself.

Newton's worldview mixed science with magic, even if, in practice and among his peers, he was extremely careful to distinguish between the two. When computing the influence of Saturn's orbit on Jupiter, there is no place for theological speculation. There is just hard-core physics and calculus. Newton insisted that his new natural philosophy was explicitly quantitative and followed a strict scientific methodology, that is, that any hypothesis had to be experimentally validated to be seriously considered as the explanation for a given observed phenomenon. An example is his famous "I do not feign hypotheses" quote from the General Scholium (a sort of epilogue) published in the second edition (1713) of his landmark book *Mathematical Principles of Natural Philosophy* (often called the *Principia*):

> *Thus far I have explained the phenomena of the heavens and of our sea by the force of gravity, but I have not yet assigned a cause for gravity ... I have not as yet been able to deduce from phenomena the reason for these properties of gravity, and I do not feign hypotheses. For whatever is not deduced from the phenomena must be called an hypothesis; and hypotheses, whether metaphysical or physical, or based on occult qualities, or mechanical, have no place in experimental philosophy ... And it is enough that gravity really exists and acts according to the laws that we set forth, and is sufficient to explain all the motions of the heavenly bodies and of our sea.*[8]

Newton was proudly aware of the power of his theory of universal gravity to explain the observed natural motions of a huge variety of gravity-based phenomena on Earth and in the skies. The theory worked. But he confessed to not knowing what gravity was. He didn't know *why* two bodies with mass would attract one another. Smartly, he decided not to guess—"I do not feign hypotheses"—and proceeded to lay down the essential empirical nature of the physical sciences: "For whatever is not deduced from the phenomena . . . [has] no place in experimental philosophy." We use the word "hypothesis" somewhat differently nowadays, but Newton meant that if you can't come up with experimental tests, your theory is worthless for science.

Sounds like a strong and clear-cut statement. Note, however, how Newton qualifies the possible kinds of hypotheses: "whether metaphysical or physical, or based on occult qualities, or mechanical." He seems to be implying that there are other modes of explanation apart from the scientific: metaphysical, occult ways of knowing. They may not have a place in "experimental philosophy," but they certainly had a place in Newton's mind. Indeed, five years after he published the *Principia*, he exchanged letters with Oxford theologian Richard Bentley, who wanted to use universal gravity to prove the existence of God. As a theist, Newton was very receptive. Bentley asked for Newton's views on the nature of gravity and got the following answer:

> *Tis unconceivable that inanimate brute matter should (without the mediation of something else which is not material) operate upon and affect other matter without mutual contact; as it must if gravitation in the sense of Epicurus [Atomism] be essential and inherent in it. And this is one reason why I desired you would not ascribe innate gravity to me. That gravity should be innate inherent and essential to matter so that one body may act upon another at a distance through a vacuum without the mediation of anything else by and through which*

their action or force may be conveyed from one another is to me so great an absurdity that I believe no man who has in philosophical matters any competent faculty of thinking can ever fall into it. Gravity must be caused by an agent acting constantly according to certain laws, but whether this agent be material or immaterial is a question I have left to the consideration of my readers.[9]

Could gravity be mediated by "something else which is not material"? Newton goes on to explain that the only way gravity could act at a distance, as it does between Sun and Earth, and still be consistent with Epicurean Atomism (objects can influence others only by colliding) was by some agent acting through space. He leaves the decision of what kind of agent that is (material or immaterial) to his readers, opening the door to magical thinking. And he likes it this way, given that elsewhere in the General Scholium Newton attributes the splendor of the cosmic order to "an intelligent and powerful Being," arguing that "No variation in things arises from blind metaphysical necessity, which must be the same always and everywhere. All the diversity of created things, each in its place and time, could only have arisen from the ideas and will of a necessarily existing being."[10]

To Bentley, Newton explains that this Being must be "very well skilled in Mechanicks and Geometry."[11] God was the Cosmic Designer, a seventeenth-century representation of Plato's Demiurge. Pleased, Bentley presents Newton with another conceptual challenge, asking how a finite and spherical Universe filled with mutually attracting chunks of matter would not collapse into a huge blob at its center. In other words, to what could we attribute the stability of the Universe as a whole? Newton then introduced a revolutionary idea. The Universe, he claimed, must be infinite. Only then, attractions from all directions would cancel each other and render the whole of Creation stable. Even more, God would act to correct any slight gravitational imbalance, such as that caused by a comet passing near a planet. To Newton, the very existence of the Universe and its endurance through time was proof of God's design and continuous presence. British economist and historian of ideas John Maynard Keynes famously wrote that "Newton was not the first of the age of reason. He was the last of the magicians."[12] Newton's science was a reverential pilgrimage of the mind to decipher God's plan for the Universe.

Newton was the dividing line between two profoundly different worldviews: one where the world is filled with magic, and one where it isn't. The spectacular success of his science seeded the so-called Age of Reason or Enlightenment, founded on a strictly materialistic and rational view of reality that dictated not just how science should be done, but how people should relate to the natural world. Newton's theism—a God present at all times—gave way to deism—an absent God whose role was relegated to that of creator of the Universe and its laws. Ironically, the very precision of Newton's science exorcised the need for God's continuous presence in the world. From then on, the business of science was to decode the clockwork mechanics built within the framework of a reality composed of material bodies and forces acting on them. After Newton, Nature lost its soul.

With no gods to protect it, Nature was desacralized to become a commodity, an object to be exploited for capital gain. In one of the greatest hypocrisies of history, the rational and civilized white man of the West crushed the "savage" and "primitive" Indigenous communities of the Americas, Africa, and the Pacific that dared to challenge the European march toward "progress." The steam-engine-powered Industrial Revolution provided the technological means for the ever more efficient and devastating exploitation of natural resources. The greater the progress, the greater the need for resources. Greed stained the land and the seas as the deafening clang of machinery silenced Nature's voice. The chasm between humans and their natural roots grew wider. What was once the sacred realm of all being became a target of economic opportunity, as "civilized man" mined and drilled, razed forests, and killed animals for food and trophy. The Enlightenment, for all its great discoveries and creations, was also the age that amplified humanity's moral failings, as it turned reason into a deadly weapon of environmental destruction.

CHAPTER FOUR

THE SEARCH FOR OTHER WORLDS

> It would be very strange that the Earth was as populated as it is, and the other planets weren't at all, for you mustn't think that we see all those who inhabit the Earth; there are as many species of invisible animals as visible.
>
> —Bernard Le Bovier de Fontenelle, *Conversations on the Plurality of Worlds*

THE VARIETIES OF WORLDS BEYOND

There were protests against Nature's objectification within European society and culture. The Romantics, for one, objected to the onslaught of reason, reinstating their deep attachment to the natural world. In England, William Wordsworth and Samuel Taylor Coleridge retreated to Somerset to immerse themselves in the landscape. "While with an eye made quiet by the power / Of harmony, and the deep power of joy, / We see into the life of things," wrote Wordsworth with reverential love for his surroundings.[1] In 1818, Mary Shelley published *Frankenstein*, a gothic meditation on the perils of pushing science beyond its ethical allowance.

To "see into the life of things" clashed hard with the massive machinery of industrialization and its appetite for natural resources. The Romantics opposed the onslaught with their artistic creations in literature, music, and painting, cultivating the feeling of the sublime within Nature, seeking to reenchant the natural world. In Germany, Beethoven, with his Moonlight Sonata and Pastoral Symphony (No. 6 in F major), aligned his music with the evocative beauty of the natural world. The same year that Shelley published *Frankenstein*, Caspar David Friedrich painted *Wanderer Above a Sea of Fog*, the iconic depiction of man's search for meaning while in deep contemplation of Nature's awesome beauty and power. As writer Robert Macfarlane noted in *Mountains of the Mind*, Friedrich's *Wanderer* "became, and has remained, the archetypical image of the mountain-climbing visionary, a figure ubiquitous in Romantic art."[2] Only up on the rarefied heights could one find solitude, away from the noisy and smelly masses confined within city walls.

We see a connection between what secular Romantics searched for in Nature and the active spirituality of Indigenous cultures across the globe and of eccentric outliers of monotheistic faiths—such as the desert fathers and mothers and their ascetic life: the call to embrace the wild, to immerse oneself within untamed Nature to bond with the divine. In their unique ways, they understood that being in Nature evoked a feeling of belonging to something bigger, opening a pathway to a spiritual connection with the Cosmos that transcended the passage of time.

To some natural philosophers, the scientific exploration of

the skies also led to romantic contemplations. In Book III of the *Principia*, Newton speculates on the generation, decay, and regeneration of cosmic matter, echoing Anaximander's ancient notion of a Nature always in flux, of worlds being born and dying in endless succession (see chapter 2). Newton's lyrical vision of the recycling of matter through stars, planets, and comets blended his mechanical science with his alchemical explorations. Gravity, the great unifier, the embodiment of God in the world, orchestrated change and transformation throughout the Cosmos:

> *And the vapors that arise from the sun and the fixed stars and the tails of comets can fall by their gravity into the atmospheres of the planets and there be condensed and converted into water and humid spirits, and then—by a slow heat—be transformed gradually into salts, sulphurs, tinctures, slime, mud, clay, sand, stones, corals, and other earthy substances.*[3]

Newton's description of the ebbing and flowing of heavenly matter expresses an organic, alchemical vision of a Cosmos undergoing constant change. Wandering comets are the messengers responsible for transferring materials from stars to planets, where they undergo chemical transformations into the substances that enable life. Gravity enables this cosmic unity, woven through the sharing of star-stuff across all planets. Stellar and cometary "vapors" cooked over "slow heat" (a reference to the slow burning of alchemical experiments) generate "earthy substances." The chemistry that is ultimately responsible for life spreads across the Cosmos. Newton's alchemically inspired vision for a mechanistic Universe suggested that life elsewhere is scientifically possible.

The ensuing refinement of Newton's mechanical legacy, combined with the rapid growth of telescopic power in the eighteenth and nineteenth centuries, led to spectacularly successful calculations that confirmed the clockwork nature of the heavens. Remarkably, it also led to the discovery of new worlds.

First was Uranus, which Sir William Herschel discovered after a series of observations started on March 13, 1781, from his home in Somerset, not far in distance and in time from Wordsworth's and Coleridge's homes.[4] Contrary to what many believe, Uranus is visible with the naked eye; however, because it is very faint and moves slowly, previous astronomers had deemed it a star, as did Herschel, initially. He then changed his mind, suggesting that the new celestial object was a comet, given that its size grew as he increased his telescope's magnification power. Stars, being too distant, would remain the same pointlike size.[5] The news spread quickly across the European astronomical community, prompting various eyes to train on the new object. Was it a comet? Maybe a new planet? With patience and consistent observation, the shape of its orbit would tell. Comets tend to have highly elliptical orbits, while planets' orbits are closer to circular.

Science is akin to detective work. We look for clues to help us solve a mystery. Opinions may diverge at first, but eventually, fresh data and careful analysis drive the community toward consensus. And so it was with the new celestial object. By 1783, Herschel wrote to Joseph Banks, president of the Royal Society:

"By the observation of the most eminent Astronomers in Europe it appears that the new star, which I had the honour of pointing out to them in March 1781, is a Primary Planet of our Solar System."[6]

Uranus was the first "new" planet, a world orbiting the Sun beyond Saturn, to be discovered with an instrument. Of course, the planet was new to our eyes only, being billions of years old like its fellow planets. After millennia of naked-eye sky-watching, scientists endowed with ever more powerful telescopes were rewriting the cosmic narrative. Giddy with the promise of his newly amplified vision, Herschel compared the heavens to a "luxuriant garden which contains the greatest variety of productions, in different flourishing beds."[7]

Stars and wandering planets were no longer the sole inhabitants of the skies. Charles Messier in France and Herschel in England compiled catalogues of "nebulae"—diffuse shining objects without the pointlike nature of stars and planets—compounding the complexity of what Nature could create with the mystery of what these creations could be. With these advances came new questions, and the increased popularity of astronomy and astronomers. King George III appointed Herschel to be his Court Astronomer, bringing him to live in Windsor with his sister and collaborator Caroline so that the royal family and guests could also look through the ever more wondrous telescopes. The largest was a gigantic 40-foot reflecting telescope, with a 49$^{1}/_{2}$-inch-diameter primary mirror, at the time (1789) the largest scientific instrument ever built.[8] The instrument attracted all kinds of illustrious visitors, including a diverse entourage of scientists, poets, and foreign dignitaries, a list that included Erasmus Darwin (Charles Darwin's grandfather) and William Blake. What other worlds were out there waiting to be discovered no one could tell. Astronomers were the new hunter-gatherers, their unexplored celestial plains the endless vastness of the night sky.

What we see of Nature is always limited by what our instruments allow us to see. As long as we nurture our curiosity and have proper funding, there is no end to this exploration. This speaks both to the power and to the limitations of science.[9] As we build more powerful instruments, we sharpen our view of the Cosmos and, by extension, of ourselves. But our success shouldn't seduce us into thinking that we can ever have a complete view of the world or that a perfectly objective view is possible. There is no God's-eye view of the world, at least not for our human eyes. We are bound to see the world from within the confines of our minds and bodies.

What we discover of the world, the amplified reality our instruments allow us to see, is always framed within our very human reach, grounded by our experience of being-in-the-world. In the realm of both the very large, from solar system astronomy to cosmology, and the very small, from microbiology to particle physics, our instruments are bridges between the visible and the invisible. They translate what is beyond our sensorial reach into images, sounds, and graphs that we then interpret with our minds. All science is a flirting with the unknown. As we learn more and our vistas of reality expand, we should recall our limitations with the humility they demand.

OTHER WORLDS, OTHER LIFE

With sharper eyes, the awesome variety of celestial worlds unfolded. Herschel equipped his telescopes with prisms and temperature sensors, discovering in 1800 the existence of an invisible form of light beyond the visible spectrum, what we now call infrared radiation. Displaying his incredible ingenuity, Herschel devised a way to measure the temperatures related to the different colors of the visible spectrum and even slightly beyond the visible, in the infrared.[10] He realized that celestial objects shine with different colors and temperatures. The Sun, for example, was one degree hotter in the infrared than in red light. Objects may shine in types of radiation that are invisible to human eyes. Herschel found that celestial objects like the Sun, stars, and nebulae emit light with different properties, thus pioneering the field of astronomical spectrophotometry and changing astronomy forever. Our current cutting-edge astronomical instrument, the James Webb Space Telescope, is an infrared telescope searching for the first stars, formed about one hundred million years after the Big Bang, and for potential signatures of life (biosignatures) in exoplanets, a topic we will address soon. Herschel would no doubt enthusiastically endorse our current interest in finding life elsewhere, given his firm conviction that life was ubiquitous across the Universe.[11]

He was not alone in this belief. The Greek satirist Lucian of Samosata, who lived in the second century CE, authored the first known tale (which he confessed was an elaborate but delightful lie) of a trip to the Moon and beyond, complete with bizarre alien creatures and interplanetary warfare. Fourteen centuries later, Kepler wrote *Somnium* (see chapter 1), his own tale of a trip to the Moon, which was published posthumously.[12] *Somnium* is a brilliant exercise in considering astronomy from the perspective of a different celestial body: the durations of days and nights, the possibility of eclipses, what Earth looks like from somewhere else. As a bonus, Kepler also speculated on the kinds of creatures that could live in such a place. As we noted in chapter 1, their struggles for survival foreshadowed some of the principles that later emerged in Darwin's theory of evolution by natural selection.

The demotion of Earth to an ordinary planet allowed for life to be possible on other worlds. In 1698, *Cosmotheoros*, by Newton's contemporary Christiaan Huygens, took the possibility of extraterrestrial life to the next level, arguing it to be an unavoidable consequence of Copernicanism. Planets have water, animals, and plants "not to be imagin'd too unlike ours . . . but not just like ours." The book's opening sentences make Huygens's position clear:

> *A Man that is of Copernicus's opinion, that this Earth of ours is a Planet, carry'd around and enlighten'd by the Sun, like the rest of them, cannot but sometimes have a fancy, that it's not improbable that the rest of the Planets have their Dress and Furniture, nay and their Inhabitants too as well as this Earth of ours.*[13]

Huygens then mentions his predecessors, including Lucian, Bruno, Kepler, and Bernard Le Bovier de Fontenelle, that "ingenious French Author of the Dialogues about the Plurality of

Worlds." In 1686, one year before Newton published his *Principia*, Fontenelle's book *Conversations on the Plurality of Worlds* brought forth the possibilities that Huygens explored in depth, the kinds of inhabitants that could exist on worlds similar to but not quite like ours, including those circling distant "fixed stars." As the quick-witted Marquise comments to the philosopher, the narrator of *Conversations*, "My imagination's overwhelmed by the infinite multitude of inhabitants on all these planets, and perplexed by the diversity one must establish among them; for I can see that Nature, since she's the enemy of repetition, will have made them all different."[14]

Both Huygens and Fontenelle believed that life abounds in the Universe, experimenting with an endless variety of shapes and forms across worlds untold. Predating Newton's vision of worlds renewed by the coming and going of comets, Fontenelle's philosopher-narrator conjectures: "The Universe could have been made in such a way that it will form new suns from time to time. Why couldn't the proper matter to make a sun, after having been dispersed in many places, reassemble at length in one certain place, and there lay the foundation of a new world?"[15]

By the eighteenth century, Copernicanism had laid its roots deep into the European mindset. The analytical tool here was scientific induction: if the solar system has planets and Earth is a planet, it's reasonable to suppose that other planets have similar properties. After all, Galileo had shown that the Moon has valleys and mountains. Why not other planets? And if the Sun is a star and there are countless stars in the sky, those stars must also have planets orbiting them and hence life, just like in our solar system.

This all seems reasonable insofar as induction makes sense.

But induction is always limited by what we know of the natural world. More information can and often does contradict it. For instance, as mentioned in chapter 2, it was widely believed in seventeenth-century Europe that all swans were white. Then Dutch explorer Willem de Vlamingh found black swans in Australia in 1697. To extrapolate from a limited data sample is dangerous.

Induction is extremely useful but must be taken with a large grain of salt. We now know, of course, that the probability of life on other planets of this solar system is extremely low, with the possible (but still unlikely) exception of subterranean Mars. Of the eight planets in the solar system, four are rocky and four are gas giants—two sets of worlds with very different properties. Rocky planets may share geological characteristics with Earth, such as having mountains and volcanoes. But each has a very different history, contingent on its location in the solar system and on a variety of variables, such as its composition, mass, number of moons, and atmosphere. When we add the possibility of life into the mix, the unknowns grow exponentially. As we will see in part III, using induction to speculate about the possibility of life elsewhere is a path fraught with pitfalls. Still, science has advanced tremendously since the eighteenth century, including our exploration of and knowledge about other worlds.

LESSONS FROM VULCAN

The ever-expanding power of telescopes made details of other worlds accessible to the human eye, further feeding wild dreams of what could exist out there. Fontenelle wisely summarized our scientific (and philosophical) enterprise as follows: "All philosophy is based on two things only: curiosity and poor eyesight... The trouble is, we want to know more than we can see."[16] The tension between our curiosity and our shortsightedness, the fact that we always want to know more (a good thing) but our instruments are always limited (unavoidably), drives our will to expand the boundaries of knowledge.

By 1847, Uranus had completed about 80 percent of a full orbit around the Sun since its discovery by Herschel in 1781 (a complete solar orbit takes eighty-four years). Astronomers mapping its path noted several anomalies that defied Newton's theory of gravity. Rather than discard the theory, as some suggested, on June 1, 1845, Urbain Le Verrier proposed to the French Academy of Sciences in Paris that another planet tugging on Uranus was the culprit. Although he wasn't the first to propose this, his calculations were faultless. He even predicted, using Newtonian mechanics, the probable position of the planet to within one degree of its actual location. (For comparison, the full Moon covers about half of a degree in the sky.) On August 31, Le Verrier presented another paper to the academy, this time computing the mass and orbit of the new planet. As no French astronomer offered to confirm his prediction, Le Verrier contacted Johann Galle at the Berlin Observatory. Galle immediately put his student Heinrich Louis d'Arrest to work. After less than an hour of searching, d'Arrest found the new object at less than one degree of Le Verrier's prediction. Two nights after, Galle and his student confirmed that the new object was indeed a planet. "The planet whose position you have signaled really exists," an astonished Galle wrote to Le Verrier.[17] For the first time in history, physical theory predicted the existence of a world yet unseen.

Le Verrier's prediction was an amazing accomplishment. Imagine using paper and pencil to calculate the existence of something no one had ever seen before. Then an observer finds it, confirming the expected properties of the new object. From mind to reality. Still, the theory is built from previous observations, as was Newton's universal gravitation. Everything we do in science begins with our experience of the world. Theorizing comes after. So, instead of "from mind to reality," we should more accurately say "from reality to mind to reality."[18] Predicting the existence of new worlds, or of new phenomena yet unobserved, is the achievement dream of a theoretical physicist. To mind come Einstein's predictions of the bending of starlight due to curved space-time, of gravitational waves, of photons as particles of light, as well as of Murray Gell-Mann's prediction of quarks and the joint prediction in the 1960s of the Higgs boson by Peter Higgs, François Englert, Robert Brout, Gerald Guralnik, Carl Hagen, and Tom Kibble, a particle that was finally discovered in 2012.

A successful prediction means that the theory has something deep to say about the nature of physical reality, as if our ideas could see beyond our senses. "The unreasonable effectiveness of mathematics in the natural sciences," as Nobel laureate Eugene Wigner once pondered, "is something bordering on the mysterious."[19] But that's only partly true. Theories are not born from a vacuum; they are the product of years of work firmly anchored on observations. If you locked ten theoretical physicists in a windowless room and asked them to devise a theory of the world outside, they would get mostly everything wrong. We can't guess the world. Everything we do starts with our (limited) experience of reality.

Theoretical success often comes with emotional baggage, which tends to include overconfidence and inflated pride. At around the same time Le Verrier proposed the existence of Neptune, he was also puzzling over Mercury's anomalous orbit. Mercury moves very slowly around the Sun in an elliptical orbit that wobbles (or precesses) like a top.[20] Despite Le Verrier's remarkable mathematical prowess, his careful calculations of Mercury's orbit wouldn't match the observations. If Neptune disturbed Uranus's orbit, why couldn't the same be happening with Mercury? In 1859, Le Verrier proposed the existence of planet Vulcan, a new world orbiting between Mercury and the Sun.

Late that same year, Le Verrier received a letter from Edmond Lescarbault, an amateur astronomer and physician, claiming to have seen the new planet passing in front of the Sun. When a planet passes in front of a star, it can be seen from afar as a tiny black dot. This phenomenon, essential for finding exoplanets, is called planetary transit. (We will have much to say about planetary transits soon.) Anyone can watch Mercury and Venus transiting in front of the Sun, a spectacular but rare sight. For the transit to be observed, the Sun and the transiting planet must align with our line of sight. I watched the transit of Mercury on November 11, 2019, and the spectacular transit of Venus on June 6, 2012. Lucky chance, the sky was covered by hazy clouds, and it was easy for me to see the black spot moving in front of the Sun with the naked eye (wearing protective lenses). Mercury's next transit will be on November 13, 2032. Mark your calendars. For Venus, unless you are a semi-immortal transhuman, or a reader in the far distant future, it will be a long wait. The next transit will be in December 2117.

An excited Le Verrier announced the existence of the new planet Vulcan to the French academy in 1860. During the next decades, astronomers around the world sighted dozens of "transits," further giving credence to Le Verrier's claim. The popular imagination bubbled with excitement. Vulcan-mania ensued. However, there is another way to confirm the existence of a planet between Earth and the Sun, which is to observe it during a total solar eclipse. When the Moon blocks the sunlight, day turns to night, and we can see stars and, with luck, Mercury and Venus next to the darkened Sun. Astronomers followed solar eclipses for years, searching for Vulcan. After many false starts, a series of detailed observations in the first decade of the twentieth century failed to see any new planet. By 1908, lack of evidence brought the mystery of Mercury's perihelium back to the forefront.[21] If not Vulcan, what caused Mercury's orbital anomaly?

The solution came in 1915, with Einstein's general theory of relativity. The advance per century of forty-three arc seconds (or Mercury's perihelium advance) was due to the curvature of space around the Sun. No need for Vulcan. Einstein's theory of relativity exorcized quite a few ghosts from physics, showing that sometimes Nature requires us to take a very different look at things to come closer to the truth. Contra Newton, gravity is not a mysterious and instantaneous action-at-a-distance, but a consequence of the curvature of space around a massive body. Gravitational perturbations travel at the speed of light, which is certainly very fast but not instantaneous. In 1905, with the special version of his theory of relativity, Einstein showed that the mysterious medium thought to be required to sustain the propagation of light waves in space, the luminiferous ether, was not needed. As with Vulcan, the top physicists of the nineteenth century were convinced that the ether existed, even if it had very bizarre properties.[22] Outer space is empty, and there is no planet between the Sun and Mercury. Vulcan now exists only in fiction, as the home planet of *Star Trek*'s Mr. Spock. Appropriately, the Vulcans are the ultimate logical humanoids, heirs to Le Verrier's mechanistic logic that led to the prediction of planet Vulcan. Being a human-Vulcan hybrid, Spock embodies the battle of reason versus emotion, a caricature of the struggles between the Enlightenment's rationalism and the Romantics' emotive sensibility of the early nineteenth century. Curious and shortsighted, we humans need both, sprinkled with a high dose of humility against hubris.

LESSONS FROM MARS

As many eyes turned to hypothetical Vulcan, others turned to Mars. In 1877, taking advantage of the proximity of the Red Planet, Italian astronomer Giovanni Schiaparelli noticed striations along the Martian surface, which he described in Italian as *canali*. To some scientists, the long depressions he observed seemed to make patterns too regular to be natural geological accidents. Excitement quickly mounted. *Canali* was translated into English as "canals," instead of channels, leading to the thought that they were artificially built. If canals crisscrossed the Martian surface, they must be the product of advanced engineering. The Suez Canal, the engineering wonder of the era, had been completed in 1869, just a few years before Schiaparelli's observations.

Hundreds of Martian canals were observed and named, even

if they were only seen through telescopes, refusing to show up in photographs taken with the same telescopes. Well-known astronomers argued that photographs require long exposure times, making them more sensitive to the blurring effects of atmospheric thermal fluctuations, as when you look at turbulent air above hot pavement. They concluded that fluctuations would erase finer details such as canals. Could the canals be the work of an old and wise civilization, built to bring water from the frozen polar caps to the parched equatorial regions where alien beings lived?

That's what the financier and amateur astronomer Percival Lowell suggested in 1895, after observing and mapping the Martian surface from his private observatory in Flagstaff, Arizona. Lowell published his careful drawings and speculations about Martian intelligent life in *Mars*, a book that became an instant sensation. If the Martians built canals, they were intelligent. If the canals were to transport water, their life-form was water-based. If they were succumbing to massive droughts, they needed to find better worlds, plentiful with water. If we could see "them," they could see us. If they were older than us, their technology would be much more advanced than ours. If we could traverse the Earth in our ships, steam engines, and balloons, they could take off to space in their spaceships. If Western civilization had forcibly colonized a substantial fraction of our planet, "they" could come here to colonize us.

In 1897, just two years after Lowell published his first book on Mars, H. G. Wells's *War of the Worlds*, the first sci-fi classic to imagine dystopic fantasies about a Martian invasion, was serialized. Wells used the Martians as a metaphor for humanity's future, at the time controlled by coexisting Western empires encroaching on each other's domains. The same way that evolutionary history has shown that intelligent species seem unlikely to peacefully coexist—and by intelligence here I mean human-level intelligence, capable of transforming raw materials into life-changing technologies—the mounting tensions between empires couldn't endure peacefully for long. Prophetically, only seventeen years after the publication of *War of the Worlds*, the Western empires erupted into the Great War. In the novel, Martian science, way more advanced than ours, had created horrible machines of mass destruction that made our weapons look like children's toys. What saved humanity was not heroism or human ingenuity, but the indiscriminate force of evolution by natural selection: "the Martians—*dead!*—slain by the putrefactive and disease bacteria against which their systems were unprepared . . . slain, after all man's devices had failed, by the humblest thing that God, in his wisdom, has put upon this Earth."[23]

Life, Wells understood, is uniquely adapted to its environment. No two planets are alike. No two planets have the same physical and chemical properties, the same geological history, the same sequence of cataclysmic and planet-changing events—be they collisions with other celestial bodies or internal disruptions such as massive volcanic activity or prolonged weather changes. This means that even if we assume that life starts with the same basic ingredients everywhere in the Universe (as we will discuss in part III), it will not evolve to be the same elsewhere. For sure, certain evolutionary traits, such as bilateral symmetry—the left-right symmetry that many animals here on Earth share that optimizes the ability to see and move around—may reappear in alien life-forms. But if life emerges and evolves on other worlds, it will surely look different from life here on Earth. The fundamental question, wide open at present, is how different.

During the 1960s and 1970s, NASA's probes from the Mariner and Viking programs photographed the Martian surface and found no hint of any large-scale canals or, for that matter, of any intelligent and technologically savvy civilization, past or present. Although by then most scientists knew this, in popular culture the myth of Little Green Men from Mars still persisted. After Mariner and Viking, it became clear that "they" would have to come from farther away. What the probes did find, and which several follow-up missions confirmed, including the *Spirit* and *Opportunity* rovers that were active until 2010 and 2018, respectively, was a rich geological history of a planet that is now a frigid desert crisscrossed with vast dry river canyons, valleys, and huge extinct volcanoes. Olympus Mons, the largest of all known volcanoes in the solar system, has a diameter comparable to that of the state of Arizona. With an elevation of sixteen miles, it's about three times as high as Mount Everest. Clearly, young Mars was a different world, with plenty of water and geological activity, and a climate more conducive to life. Many believe life could have blossomed there for a while, only to be snuffed out of existence by the onset of unbearably harsh conditions. The Martian atmosphere thinned out over time, exposing its surface to deadly solar ultraviolet radiation—the same radiation that requires we use sunblock to protect our skins. Contrary to our planet, where life managed to persist for at least 3.5 billion years, even if sometimes just barely, other worlds may have hosted life for much shorter periods. For young worlds, born recently in star-forming regions, life may not have had time to emerge yet. As we will see, despite life's resilience, it does need a host of conditions to flourish and endure in challenging environments.

As I write these lines, NASA's *Perseverance* rover is hard at work, digging into the Martian soil for signs of present and past biological activity. What kind of biological activity that would be we don't know. Sample analysis follows clues based on how we understand life on Earth. After all, life here is all we know of life anywhere. The rover "follows the water," that is, it roams around the Jezero Crater, a region where water was present in the past, to search for organic compounds that point toward life as we know it.

Curiosity drives us forward, and shortsightedness slows us down. That's how the drama of wanting to know unfolds. But humans are a persistent and resilient bunch, and we keep plowing ahead. As Tom Stoppard wrote in *Arcadia*, "It's wanting to know that makes us matter."

Life can, and probably will, most certainly surprise us. But we must start from somewhere, and the obvious starting point is to use what we know of life here to look for life elsewhere. If we do find life elsewhere, and it is similar in genetics and morphology to terrestrial life, we will learn something. If it is different, we will learn something. And if we don't find it, we will also learn something, although we must always be very careful with inductive reasoning when we draw conclusions both from evidence that is found and from evidence that is absent. Indeed, if our current and future missions fail to find life on Mars—past or present—it doesn't mean that there isn't or wasn't any life on Mars. That would be a conclusion with a finality that our limited collected evidence wouldn't justify. We could have been looking in the wrong places or with the wrong mindset. A planet is a big place, and our tools of exploration are limited in their range and precision. They are also necessarily attuned to search for life as we know it. Still, how truly extraordinary it is to live at a time when we can send a probe to dig into the soil of another planet in search of life and have samples eventually shipped back to Earth to be studied.[24]

LESSONS FROM OUR SOLAR SYSTEM: WORLDS OF WONDER AND MYSTERY

There is, however, a deeper lesson to be learned, and it is not about life elsewhere. The lesson is about life here. The past sixty years of space exploration have changed the way we understand our solar neighborhood. We have explored all the planets of the solar system, and many of their amazing and bizarre moons. Although so far we have landed only on the Moon, Mars, Venus, and Titan (and some asteroids and comets), we have sent probes that fly by every planet. The list even includes faraway Pluto, now labeled a "dwarf planet." Why a dwarf planet and not just a planet? Because Pluto didn't clean up the debris accumulated along its orbit during its formation, as worlds currently defined as planets must do. In fact, Pluto fits best the description of a large member of the Kuiper belt—an outlier region beyond Neptune's orbit consisting of frozen dusty ice-balls left over from the formation of the solar system 4.5 billion years ago. Had Pluto accreted much of this debris as it circled the Sun, then it would fit the "planet" description better.[25]

Each planet is its own world, with very particular properties.

We see this easily in our solar system, comparing, say, hellish Venus with frigid Mars and, even more dramatically, with frozen gas giant Saturn. The variety is so staggering that astronomers now talk of comparative planetology, a subdiscipline devoted to comparing and cataloguing the different kinds of worlds and their properties. Herschel, as we have seen, was quite prescient when he deemed the heavens "a garden which contains the greatest variety of productions." This is the age of celestial botany. We expect most stars in a galaxy to have at least one planet, probably more. Given that there are between one hundred billion and four hundred billion stars in the Milky Way, our galaxy alone should have hundreds of billions, perhaps more than one trillion (10^{12}—a one followed by twelve zeros), planets. Add the moons, many of which could also harbor life, and the number of worlds grows to several trillion.[26] Just think: Jupiter alone has seventy-nine moons at the time of writing (the number keeps increasing). That's a thousand billion worlds or more in our galaxy alone—each different, each with its own history and geophysical properties, its own interior composition, with or without an atmosphere, with or without a magnetic field, rocky like Earth or gaseous like Jupiter. There are planets without a moon, or with one, two, or dozens of moons. There are planets and moons with active volcanoes and geysers, surface or subsurface water, and clouds that generate massive storms and hurricanes. There are completely barren worlds and, possibly, as many hope, worlds teeming with life.

Broadly speaking, planets can be rocky or gaseous, with rocky worlds usually expected to orbit closer to their host stars. I say "usually" because thousands of gas giants have been observed orbiting remarkably close to their host stars. When stars and their planets are born, the hot radiation from the host star blows off volatile gaseous materials from planets that are closer, explaining why rocky planets like Earth tend to trace tighter orbits around their stars. In our solar system at least, the four rocky planets—Mercury, Venus, Earth, and Mars—orbit closest to the Sun. The gas planets—Jupiter, Saturn, Uranus, and Neptune—are farther away. The boundary between rocky and gas worlds is the asteroid belt between Mars and Jupiter, a collection of rocky asteroids that never coalesced into a planet or moon, a leftover pile of planet-making stuff. But contrary to many sci-fi films or video games, the asteroids in the belt don't make for a crazy obstacle course. If they were collected into a single world, their joint mass would be roughly equivalent to half of our Moon's mass.

This all seems to make sense, except that reality is much more nuanced and fascinating. We don't know how typical our solar system is. What's "typical" in this context, anyway? To define typicality, we need large amounts of data. When comparing members of a data sample, "typical" is what's close to the average. To have fair data to decide whether something is typical (or anomalous), we need an unbiased, or close to unbiased, dataset. That is, we need to be able to collect data that don't drive the sample toward one end of what's possible, to the detriment of other possibilities. For example, if you want to estimate the average life expectancy of humans, you shouldn't collect data only from Asia. You need a fair sample that covers a representative number of countries around the globe. You can then compare different subgroups (different genders, Asians with Europeans or with South Americans, etc.). To have a fair-sampled dataset you need broad access to data. That's much easier to accomplish in the case of humans on Earth than in the case of other solar systems that are dozens, hundreds, or thousands of light-years away.

Inductive reasoning and typicality are often brought together, sometimes carelessly. The sentence "all swans are white," as we have seen, was the result of biased inductive thinking due to limited sampling. European swans are not a representative sample of the swans that exist in the world, only of swans native to Europe. All we can say with confidence from data collected only in Europe is that a typical swan *native to Europe* is white. But we can't say that on Earth a typical swan is white.

We can now circle back to a previous discussion and use similar reasoning to critique one of the widespread misinterpretations of Copernicanism, the assertion that Earth is a "typical" planet. Copernicus's proposal to displace Earth from the center of the solar system did not (and could not) make any statements about Earth being or not being a "typical" planet. All that Copernicus did, and correctly of course, was to propose that Earth is a planet. There was nothing in his proposal adding "typical" to Earth as a planet. Indeed, typicality cannot be defined in our solar system. There aren't enough rocky or gas planets to define what's typical. If we were to limit our comparison to either the rocky or the gas planet subsets, we would clearly see that none of the four rocky planets can be defined as typical given that they are so different from each other. The same with the four gas planets. So, if one insists on defining a typical planet, one would need a large sample of planets orbiting other stars. And even then, stars are different. They have different sizes, masses, surface temperatures, and radiation outputs. Different stars affect planets differently. Thus, a fair comparison between planets could either restrict itself to planets orbiting the same kind of star (more about that soon) or to planets orbiting within what we define as the *habitable zone* of a star.

Broadly speaking, the habitable zone of a star is the belt-shaped region around the star that demarcates the area where a planet could have liquid surface water.[27] In short, a planet in the habitable zone is a potential host for life. This zone is sometimes called the Goldilocks zone, where the planetary surface is not too hot for water to evaporate and not too cold for it to be frozen. Although habitable zones are a useful first step in searching for watery rocky worlds, defining habitable zones is a complicated business. Many subtleties come into play, from the position of the host star in the galaxy—a galactic habitable zone concept—to details of the planet's chemical and atmospheric composition, to whether surface water is even a decisive factor for life (it isn't).

For example, a planet may be in the habitable zone of its host star but may be exposed to too much radiation (for example, from nearby supernovae) or have chemistry not conducive to life. Or, like with Jupiter's moon Europa, which is far from the Sun's habitable zone, liquid water may be hidden under a frozen icy crust about one or two miles thick. The current consensus is that a sixty-mile-deep water shell surrounds Europa's rock and iron-rich core. That's an ocean about ten times deeper than the deepest trenches of our own oceans, and with at least twice the volume of water of all terrestrial oceans combined. What makes Europa's subsurface water liquid is not the Sun's heat but Jupiter's gravitational pull, which is so intense that it flexes the moon's interior like Play-Doh as it moves along its elliptical orbit, transforming mechanical energy into heat, a phenomenon called tidal heating. Io, the moon closest to Jupiter, has at least four hundred active volcanoes, with extreme tidal heating making it the most geologically active body in the solar system. In Io's case, the subsurface ocean is a mixture of solid and molten rock that relentlessly bubbles up to the surface to relieve underground pressure. The first time I saw pictures from the *Galileo* space probe of Io's tortured surface covered with lava fields and erupting volcanoes, I couldn't help but remember the drawings of one of the volcanic worlds in *The Little Prince*. Reality is definitely stranger than fiction.

Venturing from Jupiter's moon system to Saturn's, we find another marvel of geological mischief, Enceladus, the sixth largest of Saturn's currently known eighty-three moons. (Only sixty-three are confirmed at the time of writing.) William Herschel discovered Enceladus on August 28, 1789, using his brand new forty-foot telescope, having no clue of the many wonders that this new member of the celestial garden hid from distant human eyes. In 2005, NASA's *Cassini* probe flew close enough to collect samples of materials jetting out of Enceladus's surface. As with Jupiter's Europa, Enceladus undergoes tidal heating as it orbits Saturn, having a warm enough interior to feed spectacular volcanic and geyser activity. A cratered and relatively quiet north pole contrasts with a hyperactive south pole where at least a hundred "cryovolcanoes" shoot up geyserlike jets of water vapor, salt crystals (sodium chloride), ammonia, ice particles, and other solid materials at a rate of about 440 pounds per second. Some of it falls back to the surface as a kind of snow, while the rest makes up most of one of Saturn's rings, known as the E ring, a broad structure spanning about twelve hundred miles between the moons Mimas and Titan, Saturn's largest.[28] Matter from the insides of Enceladus cycles back into other worlds, embodying Newton's alchemical vision of death and rebirth across the Cosmos.[29]

To add to Enceladus's remarkable properties, scientists conjecture that, like Europa, it also hides a subsurface ocean, a trapped water pocket rich in methane with an estimated depth between sixteen and nineteen miles, about four times deeper than our terrestrial oceans. Furthermore, since salty water spews out of the cryovolcanoes, this subsurface ocean is probably salty with added simple organic compounds, suggesting the possibility that it contains some of the basic building blocks for life. The combination of a salty subsurface ocean, an active hydrothermal cycle that results in volcanic activity and the circulation of materials, and the presence of ammonia and complex organic compounds makes Enceladus a prime target to study an alien environment potentially conducive for life. Several missions have recently been proposed to search for traces of biological activity in this strange world.

Here on Earth, ancient microorganisms dating back to the first traces of life more than three billion years ago combined hydrogen and carbon dioxide to obtain energy, generating methane as a byproduct. Scientists currently believe that this reaction, known as *methanogenesis*, is at the root of the tree of life on Earth. Enceladus seems to have the right ingredients for similar microorganisms to have emerged there. Although we will only know by looking—and let's hope exploratory missions are funded soon so that our curious and shortsighted human eyes can investigate this cold world—the possibility is certainly tantalizing.

But even if we do look and find no trace of microscopic life on either Europa or Enceladus, or on any other world in our solar system, two lessons are already clear: first, that our world, far from being a "typical" planet, is the true wonder of the solar system, with a thriving biosphere bursting with creatures of astonishing variety of function and form; and second, if worlds in our solar neighborhood are so diverse, imagine what marvels await us as we venture beyond, to the distant wilderness of other stars and their orbiting worlds.

HOW TO FIND NEW WORLDS 1: LOOK FOR STARS

Going beyond the solar system we encounter two major obstacles. The first is the huge distance between stars. For a rough estimate, the trip to Alpha Centauri, the nearest star system, at 4.37 light-years away from the Sun, would take our fastest spaceships roughly one hundred thousand years. We obviously can't send probes there to get useful information. The second obstacle is that our current telescopes can't directly detect planets orbiting other stars. To study exoplanets, we need different approaches that can still reveal some properties of these distant worlds. During the past decades, remarkable feats of ingenuity and technological advances have allowed us to explore thousands of these worlds. And the news is truly spectacular.

After centuries of speculation, we now know that planets are orbiting pretty much every star in our galaxy. We should revisit the astonishing numbers again. There are an estimated one hundred billion to four hundred billion stars in the Milky Way. If every star has an average of one to five planets, the numbers range from one hundred billion to two trillion exoplanets. And then there are the moons, which, as we have learned from our own solar system, no doubt hide amazing surprises. Including planets and moons, we can comfortably estimate that there are trillions of worlds in our galaxy alone, each different, each with its own history, geophysical properties, and composition. Herschel's celestial garden has far surpassed everyone's expectations.

What kinds of worlds are these? How many are similar to Earth? How many can host life? How many do? Is our solar system "typical," or are we, especially Earth, the odd world out there, pulsating with life in a dead Universe? In the coming decades we should be able to answer most if not all of these questions. But I would argue that we have already learned enough to position Earth as at least an exuberant oddity among planets. How unique we are as a technologically savvy civilization in the Cosmos we cannot say with certainty; however, we can say that if there are others, they have been quite shy about making contact. Maybe they have similar technological obstacles reaching us as we have reaching them. In any case, and this is a central point for us, what we have learned about life on our planet and the solar system is enough to realign our thinking about who we are in the Universe and why we have a leading role to play in the unfolding cosmological narrative.

This leading cosmic role for humanity is not due to the dictum attributed to the Greek pre-Socratic philosopher Protagoras of Abdera, who famously proclaimed that "Man is the measure of all things," as we most certainly are not. However, we *are* the things that can measure. If there is anything special about our species, it is not that each of us is capable of deciding what truth is, as Protagoras believed (a position that Plato greatly disliked since it implied a relativism that made ultimate truth impossible to attain), but that we are sentient beings capable of building devices that can expand our view of reality, allowing us to discover our place among the stars.

The story of how modern astronomers discovered new worlds beyond our solar system has been told in many excellent books.[30] For us, it's enough to briefly survey the methods used and what we have learned so far about exoplanets and their properties.

We begin with the kinds of stars that exist and whether their orbiting planets can host life. There are seven kinds of stars that shine by fusing hydrogen into helium at their cores. The differences have to do with how big the star is (how massive compared with the Sun) and how efficiently it burns its fuel. They range from giants that have masses sixty times larger than the Sun (called type O stars, or blue super giants) to the smallest stars, known as type M stars—aka red dwarfs—with about one-fifth the mass of the Sun.[31] Note the "blue" and "red" attached to the names of the two extreme kinds of stars. Blue indicates very high surface temperature, and red, low surface temperature.[32] The rule of thumb is that the more massive the star, the hotter it is. Its larger gravity squeezes matter at its core harder, increasing the efficiency of the fusion process. Stars self-cannibalize to fight the inexorable pull of gravity that tries to implode them. The harder they fight back, the stronger they shine.

A star's temperature matters for finding life on its orbiting planets or moons. If a star is too hot, it will emit too much radiation, including deadly ultraviolet rays. So, planets with orbits too close to such stars would have no chance of harboring life. Hot stars have a faraway habitable zone. The hotter the star, the farther the habitable zone. Unless, of course, life is subsurface, and thus shielded from the deadly rays. We have discussed this possibility with Europa and Enceladus and their huge oceans hidden under an ice crust.

There is also the problem of the star's longevity. The more massive the star, the shorter it lives. Type O stars have a life expectancy of only about five hundred thousand years, not long enough for life to emerge and flourish, as we will see. With their huge masses accelerating the nuclear fusion process at their cores, type O shine the brightest and die the youngest. In the game of life, smaller, longer-living stars have a distinct advantage.

As we mentioned, from type O to type M, there are seven kinds of stars. In order of mass (and temperature) from heaviest (hottest) to lightest (coolest), the sequence reads O-B-A-F-G-K-M. We memorize this with a mnemonic from less politically correct times in astronomy: O Be A Fine Girl Kiss Me. The following table summarizes the various stellar properties, each of them with direct implications for finding life on a star's orbiting planets and moons.[33]

STAR TYPE	PERCENTAGE IN THE GALAXY	SURFACE TEMPERATURE (°C)	LUMINOSITY (SOLAR UNITS)	MASS (SOLAR UNITS)	LIFETIME (YEARS)
O	0.001%	50,000	1,000,000	60	500,000
B	0.1%	15,000	1,000	6	50 million
A	1%	8,000	20	2	1 billion
F	2%	6,500	7	1.5	2 billion
G (Sun)	7%	5,500	1	1	10 billion
K	15%	4,000	0.3	0.7	20 billion
M	75%	3,000	0.003	0.2	600 billion

Let's spend some time going over this table because knowing the kinds of stars is essential to searching for life in their neighborhoods. "Solar units" means compared with the Sun. For example, a type B star with six solar mass units has a mass six times that of the Sun. Our Sun is a type G star. The first column, "Star Type," lists the seven different types. The second column, "Percentage in the Galaxy" gives the approximate percentage of stars of that type in the Milky Way. We see that type O stars are extremely rare, only one in one hundred thousand. Stars like our Sun are only 7 percent of the total. By far, the most abundant star type is M, a whopping 75 percent. Three out of four stars in the galaxy are type M, the little cold red dwarfs. Still, keep in mind that there are at least one hundred billion stars in the galaxy; so even if type O stars are much rarer, there are still about one million of them out there.

The table also shows the surface temperature, luminosity, and mass of each star type. A star's luminosity measures how much radiation, visible and not visible, it emits per second. There is a huge change as you go from top (type O) to bottom (type M), from very hot and bright to very cold and dim stars, compared with the Sun. Life as we know it needs warmth and can grow and reproduce only within a very narrow temperature range. Even if we include exotic life-forms called extremophiles, capable of surviving temperatures hotter than boiling water and extreme cold, the range is between -15ºC and 122ºC (5ºF and 251.6ºF).[34] This range is important for determining the habitable zone for a star type, that is, the region where life would be possible at the surface. Clearly, O-type stars have very faraway habitable zones, while M-type stars have habitable zones that are quite near.

This means that planets that orbit close to types O, B, and A stars have little chance of sustaining life (too much heat and radiation). The same goes for planets that orbit too far from types K and M stars (too cold).

The last column of the table shows the lifetimes of the different types of stars in years. A star's longevity is deeply related to the possibility of life on one of its orbiting worlds. Recall that massive hot stars live short lives, while cooler stars live much longer. For comparison, our Universe has been around for 13.8 billion years, the time since the Big Bang. The Sun is about 5 billion years old, so it's a middle-aged G-type star. In about 5 billion years, it will inflate into a red giant, swallowing Mercury and Venus and snuffing out life on Earth, if any still exists by then.

The host star determines the possibility and duration of life on its orbiting planets and moons. Life takes time to emerge and disseminate across a planet, since it requires not only the proper chemistry but also a relatively calm environment to reproduce and spread out. On Earth, the only example we know, life took at least five hundred million years to emerge and spread across the surface. More conservative and reliable estimates place life's origin at one billion years after Earth formed.[35] If life emerged before then, heavy volcanic activity, combined with incessant and devastating collisions with asteroids and comets, created a forbidding environment for life to thrive.

Still, life could have had many failed attempts to emerge on Earth before it finally managed to take hold and spread about 3.5 billion years ago (or possibly earlier). Whether this is the case is very difficult to determine. Given that the memory of Earth's past is registered in rocks, we know very little of our planet's early infancy. Just as we have no detailed recollections of our babyhood experiences because of the lack of neuronal substrate to support long-term memories before a certain age (about three years old), the infancy of our planet is lost in rocks that were churned and molten multiple times before solidifying into a crust. Unfortunately, for life's early days there were no parents making videos or taking photographs. Life's origins and earliest steps are lost in unknowable shadows.

HOW TO FIND NEW WORLDS 2: LOOK FOR PLANETS

Having determined which stars are the most promising hosts for planets capable of bearing life, the next step is to find planets orbiting them. Here is where things get complicated. Having lived under a star all your life, you know that stars are very bright. Planets and moons, on the other hand, are much smaller than their host stars and shine only because they reflect the star's light. To make everything even harder, stars are extremely far away. They look, as we know, like little dim points of light. Any planets orbiting around them are much dimmer. This means that even our current largest telescopes can't quite image planets orbiting other stars with the required resolution.[36]

For these reasons, astronomers need different techniques to find exoplanets. Fortunately, planets affect their stars in subtle ways that we can measure. Imagine you go on a hike with a friend. He's a fast hiker and is way ahead of you. At some point, your friend starts swatting frantically at something. You can't see what's afflicting him, but you know from his movements that he's being attacked by something tiny and annoying. You deduce flies are the culprits and quickly put on your face net. You also know that the more frantic the swatting, the closer the flies are to his face.

Stars are not attacked by their planets, but they do react to their presence in at least two ways: first, planets tug at them gravitationally, making them wobble; and second, a planet blocks a tiny fraction of the star's light when passing in front of it. The bigger the planet and the closer it is to the star, the larger these effects are. So, instead of trying to image the planet directly, we measure the effects planets have on their host stars, either by making them wobble (the *radial-velocity* [or *Doppler*] *method*) or by dimming their light (the *transit method*). We then work backward to deduce the kinds of planets that could cause such effects on their host stars. These indirect imaging techniques require incredible precision and care. But they work, and phenomenally well.

Technique 1: Radial-Velocity (or Doppler) Method

The first technique uses the fact that stars wobble—very subtly, but they wobble. They may look fixed to our human eyes, but under powerful telescopes we can capture their shy dance. Gravity is a tug-of-war between any two or more bodies with mass. Even if a star has much more mass than a planet, it will still feel and react to the planet's attraction. (In fact, they pull on each other the same way, as Newton's third law of motion, the law of action and reaction, tells us.) The important concept here is the center of mass. If two bodies have the same mass, the center of mass of this two-body system is exactly at the midpoint between the two. Someone sitting at this midpoint won't feel pulled in either direction, since the equal opposite pulls cancel each other. If the two bodies spin around each other, they will spin about this midpoint. The more massive one of the two bodies, the closer the center of mass is to it. When one body is much more massive than the other—like a star and a planet—the center of mass is almost at the center of the more massive body. But not quite.

For example, the Sun is about one thousand times more massive than Jupiter. If you forget about all the other planets in the solar system (a good first approximation since Jupiter is by far the most massive planet), the center of mass of the Sun-Jupiter system is at one-thousandth of the Sun-Jupiter distance, a point that lies just outside the solar surface. Both Jupiter and the Sun complete one orbit around this center of mass point in about twelve years. It's a large orbit for Jupiter and a small one for the Sun. But the Sun wobbles. An extraterrestrial astronomer observing our solar system from far away could detect the existence of Jupiter and determine its mass just by examining how the Sun moves about. No need to find Jupiter at all. Adding other planets will complicate the shape of the wobbling motion, but the physics is the same. With patience and precise measurements, it is possible to deduce the existence and masses of all planets in the solar system just by observing the Sun dance. Of course, with eight planets it gets very complicated. But it is possible.

In practice, what astronomers observe is not really the wobbling dance of the star but how its light varies as it moves about the center of mass. When a light source moves toward or away from us (or when we move toward or away from a light source), its light changes. This extraordinary effect is absolutely essential in astronomy, as it allows astronomers to measure how fast things move away or toward us, from wobbling stars to the expanding Universe. (We measure the component of the wobbling velocity in our direction, known as the *radial velocity*, hence the name radial-velocity method.)

The idea that motion affects waves was first demonstrated for sound waves by Austrian physicist Christian Doppler in 1842. Most people are familiar with this effect. When you stand on a sidewalk and an ambulance drives by blasting its siren, you notice that as the ambulance approaches you, the siren's pitch goes up, and as it speeds away from you, the pitch goes down. The same happens with cars, trucks, and trains blaring their horns. Since there were no cars yet in the 1840s, the meteorologist C. H. D. Buys Ballot tested the Doppler effect in 1845 by setting musicians on a train and asking them to play the same note, a G. Buys Ballot and a group of musical experts were positioned along the tracks to discern subtle changes in pitch. The train would ride by at maximum speed while the musicians played the same note together. The experts would then tell Doppler how the pitch changed as the train passed by them. Buys Ballot verified Doppler's formula by describing how the frequency of the note changed with the speed of the train. It must have been a fun experiment to witness.

The same happens with light waves. Waves in this sense don't mean waves crashing at the beach, but waves like the ones you see when you throw a stone into a lake. The distance between the wave crests is called the *wavelength*. Long wavelengths mean waves with crests that are far apart, while short wavelengths mean waves with crests that are closer to one another. The frequency of a wave is simply the number of wave crests that pass by a point in one second. So, if two wave crests pass by you in one second, the wave has a frequency of two cycles per second, or two hertz. When you listen to a radio and the announcer says, "Radio River, ninety-eight megahertz," it means that that radio station broadcasts using radio waves with a frequency of ninety-eight million cycles per second, or ninety-eight megahertz (MHz).

Whenever I think of waves, I have memories of my father playing his beloved accordion, an old Scandalli that had been in my family for generations. I wasn't older than five then, watching in wonder as my father played that strange contraption, stomping his feet on the ground to mark the tempo. As he opened and closed his arms to the rhythm of a song, the bellows expanded and contracted, like waves. I could never have imagined then that my father's magical music-making would one day help me understand the physics of the stars.

The Doppler effect proved that when the source of light waves (be it a laser or a star) is moving toward you, the waves get "compressed" in the direction of motion, and this compression implies shorter wavelengths and thus higher frequencies. When the light source moves away, the waves get stretched to lower frequencies. That's why the Doppler effect is so important for astronomy. It tells us whether a celestial object—be it a star, a galaxy, or a cluster of galaxies—is approaching or receding from us and how fast. If it's approaching, the light is said to shift to the blue; if it's receding, the light shifts to the red.[37]

Ground-based and space telescopes like the Hubble and now the James Webb scan the skies looking for stars with a measurable rhythmic succession of blueshifts and redshifts toward and away from us. Once they find a candidate, the repeating pattern of shifting blue-to-red-to-blue light is carefully tracked to estimate the masses and distances of the planets from their host star. Of course, the more massive the planets and the closer they are to their host stars, the more the star wobbles. As a consequence, this method works best for very massive planets orbiting close to their host stars, since they will then cause a more noticeable wobble and thus a more noticeable Doppler shift. This shift in the star light is then translated into a shift in the velocity of the star as it moves back and forth about the center of mass.[38]

In 1995, a planet was discovered orbiting the star 51 Pegasi by measuring the star's rhythmic wobble that repeated every four days with a velocity of about 187 feet per second. This was the first exoplanet found orbiting a Sunlike star using the Doppler method (51 Pegasi is a G-type star with temperature and luminosity similar to those of our Sun). Swiss astronomers Michel Mayor and Didier Queloz won the 2019 Nobel Prize in Physics for this remarkable discovery. The planet, now called Dimidium, is known as a *hot Jupiter*, a gas giant orbiting very near its host star. To everyone's surprise, Dimidium orbits 51 Pegasi closer than Mercury does our Sun. As can be imagined, this result caused tremendous excitement in the astronomical community—not only for being the first discovery of an exoplanet using the Doppler method, but for forcing us to rethink what planetary systems look like. No one would have guessed that gas giant planets could orbit this close to their host stars. For contrast, Jupiter's orbit around the Sun takes approximately twelve years. Compare that with Dimidium's four days!

This game-changing discovery begs the question: Is our solar system—with gas giant planets orbiting the farthest from the Sun—the rule or the exception? We return to the notion of typicality, but now with our solar system occupying center stage. How typical is our solar system among the hundreds of billions of other planetary systems in our galaxy?

The discovery of other planetary systems with giant planets orbiting near their host star means that it's a dangerous (and incorrect) strategy to use inductive thinking to conclude what a "typical" planetary system looks like. Only a very large sample of planetary systems could tell us more about what a typical system is, or more important, whether it even makes sense to define a typical planetary system. For example, all humans on Earth belong to the same species. This means we share many common traits, from the shapes of our bodies to our genetic makeup. But can we define a typical human? Not really. The staggering diversity of humans on this planet is nothing short of remarkable. We can't (and shouldn't) point to a single human and say that that person is a typical human. Each one of us is the product of a complex and unique confluence of environmental (phenotypical) and genetic factors. Likewise, even if planetary systems emerge from the same fundamental laws of physics and from naturally occurring rosters of chemical elements (maybe we can call those a sort of planetary genotype), each evolved according to local variations and details that will, in the end, result in a unique planetary family, with so many planets, some rocky and others gaseous, some orbiting closer to their host star (or stars) and others farther away, none equal to another. There might, of course, be general patterns that repeat across the Cosmos as we search for different planetary systems, but within each of these families of planetary systems that share similar traits, the details will be unique to each system belonging to that family. For example, there may be a family of planetary systems that has rocky planets orbiting closest to their host star, followed by gaseous planets. This family would include our solar system as well as others with a similar pattern, even if no two planetary systems belonging to this family would be exactly alike. And then, within that system, each world would have a story of its own to tell.

From 1995 until the launching of NASA's extraordinarily successful Kepler mission in 2009, the Doppler shift or radial-velocity method reigned supreme as the preeminent exoplanet finder, with about a thousand worlds to its credit. As we have seen, its success relies on planetary systems having large planets orbiting near their host stars so as to generate a Doppler shift measurable from Earth. Smaller mass host stars are better since they are tugged more dramatically by their orbiting planets. The resulting population of discovered exoplanets, not surprisingly, is biased toward hot Jupiters in short orbits around their host stars, many of them M-type stars—not very relevant as candidates for Earthlike planets orbiting a G-type star like our Sun. However, the technique not only allows for a highly successful systematic search for exoplanets, but also highlights the confounding diversity of planetary systems. Although still very much in use today, this technique has been mostly overshadowed by the transit method. When the two are used together, however, the existence of an exoplanet can not only be confirmed, but its mass and radius (and hence composition as rocky or gaseous) can be accurately estimated. This is what most astronomers mean when they call a planet Earthlike: a planet having mass and radius similar to those of Earth. Clearly, this is a qualitative criterion that says nothing about the possibility that such a world holds life, although if the planet orbits within its host star's habitable zone, the chances for life increase considerably. Still, life requires much more than these astronomical preconditions.

Technique 2: Transit Method

Naming the first space telescope designed to find exoplanets using the transit method after the extraordinary seventeenth-century German astronomer Johannes Kepler couldn't have been more appropriate. Using his brand-new laws of planetary motion, Kepler was the first in history to predict the 1631 transits of Mercury and Venus. As we have seen, a planetary transit denotes the periodic passing of a planet in front of its host star. A well-positioned observer will see a tiny black dot moving slowly in front of the star. I felt a lump in my throat when I witnessed the transit of Venus in 2012. There, for everyone to see (with proper eye-protecting filters), was an indisputable demonstration of the power of human thought to decipher, even if incompletely, the wonders of the natural world. It's hard to witness such an event, or a solar eclipse, and not feel a deep connection to the Cosmos, to what Einstein called "the mysterious." The event is a concrete astronomical phenomenon, and many may be content considering it just as such. But why push away the emotive power of the experience? Witnessing an alien world passing in front of our host star moves us in tangible and intangible ways. We see both with our eyes and with our hearts, a combination so unique to our species. There is an expansive power to the experience, if only we open ourselves to it. When we look at the Universe, the Universe looks back at us. We, alone, are aware of this. When awe colors what we see, reality becomes more magical.

Almost four centuries before my experience, Kepler predicted that Mercury would pass in front of the Sun on November 7, 1631, followed by Venus on December 6. Kepler's calculations were truly remarkable, proving once again that the motions of the planets around their host star follow simple yet precise mathematical laws. The Cosmos, at least at the level of planetary motions, is indeed a giant clockwork. If you know the laws that make the clock tick, as Kepler did, you can predict planetary transits, solar and lunar eclipses, and even the return of comets, as Newton did a few decades later.

Tragically, Kepler died in 1630, one year before he could witness the triumph of his visionary new science of the skies. In his turbulent life, tragedy followed him like his shadow, never abating for very long. On a cold morning in early November, frail and destitute, one of the greatest minds ever to walk this world went off alone on a mangy horse after patrons who owed him money. Caught in a frigid snowstorm, Kepler persisted on his quest, despite the wind and cold. He died on November 15, delirious with high fever, frantically pointing to his head and to the sky. Scattered during the Thirty Years' War, his remains were lost forever. His epitaph, which he had composed years before, is a moving expression of his love for astronomy, in particular an astronomy based on measurement:

I measured the skies, now I measure the earth's shadows.
Skybound was the mind, earthbound the body rests.

Kepler remains celebrated to this day as the pioneer of modern mathematical astronomy. NASA's mission, planned to collect data for fewer than four years, lasted more than nine, finding a staggering 2,708 confirmed exoplanets using the transit method, forever changing our understanding of planetary systems. The satellite was officially decommissioned on November 15, 2018, the 388th anniversary of Kepler's death. Its remains floating in the darkness of space, trailing Earth in its solar orbit for the foreseeable future.

During ten nights in August and September 1999, a team led by American astronomer David Charbonneau used a 3.94-inch-diameter telescope equipped with a highly sensitive charge-coupled device (CCD) camera to follow the transit of HD 209458 b, an exoplanet that had just been discovered using the radial-velocity (Doppler) method. (HD 209458 is the Sunlike host star of this planetary system.) The planet was identified as a gas giant with a radius about 25 percent larger than that of Jupiter, but significantly less massive.[39]

The method was an immediate success, quickly leading to further searches and developments. Since astronomers know the star type from its spectrum, transit alone allows for determining the planet's diameter. A large planet will dim the star more than a smaller one. Combined with the radial-velocity method—as Charbonneau and his collaborators did—astronomers can then estimate the planet's mass. Once the planet's size and mass are known, the next step is to calculate its density, that is, how much mass per volume it has. From the planet's density, astronomers can infer whether it is rocky like Earth, gaseous like Jupiter, or somewhere in between. We can infer the properties of worlds hundreds of light-years away by watching how they make their host stars dance.

The difficulty of the transit method is that for us to observe the planet passing in front of its host star, the orbit must be almost exactly edge-on with respect to Earth. Think of seeing a moth flying around a streetlamp. Of all the moth's random circling paths around the lamp, you will notice only some of the light being blocked when the moth passes between the lamp and your eyes. As you move farther away from the streetlamp, both the lamp and the moth get smaller. Only edge-on passes, or close ones, will be noticeable.

Finding exoplanets with orbits aligned just the right way so that we can spot them from light-years away requires some luck. Fortunately, the technology used for detecting transits can easily overcome this limitation: the highly sensitive cameras can look at tens, even hundreds of thousands of stars simultaneously, targeting the ones with a subtle periodic dimming, the telltale sign of a planetary transit. So, the small odds of finding a transiting exoplanet aligned for us to see are offset by looking at a huge number of stars at the same time. This is what the Kepler mission did. The Transiting Exoplanet Survey Satellite (TESS) followed, launched in April 2018.

TESS is another success story, now in extended use after completing its two-year primary mission in 2020. By early 2023, it had identified 6,176 exoplanet candidates, of which 291 had been confirmed. (Every exoplanet that is first identified using radial-velocity or transit techniques must be confirmed using ground-based, as opposed to satellite, telescopes.)

Putting together the results gathered from all satellite missions and ground-based observations, we find that so far the vast majority of exoplanets are more massive and larger than Earth.[40] Many are hot Jupiters, huge gas giants that orbit very close to their host stars. Neptunelike and larger gas giants dominate the current charts, with 3,470 out of 5,272 total confirmed exoplanets from all satellite missions and other approaches. Exoplanets that do have mass and radius closer to those of Earth tend to have much shorter orbital periods: instead of 365 days (one year) orbiting their host star, they mostly cluster between 1 and about 60 days. (For comparison, Mercury's orbit takes 88 days.) Most of these exoplanets are called *super Earths* because they are more massive and larger than our planet. They orbit substantially closer to their host stars and thus are exposed to large amounts of radiation. There are currently 1,602 confirmed super Earths. Unless their host stars are much cooler than the Sun, these planets will be too hot to sustain life, at least on the surface. They will probably be tidally locked, that is, they will not rotate about their axis, always showing the same side of their face to their host star. Or they will have a near tidally locked resonance, thus rotating very slowly about their axis, like Mercury.[41] Such worlds are not very hospitable for life.

Of the 5,272 confirmed exoplanets, 195 (or 3.7 percent) are terrestrial, meaning they have mass and radius close to Earth's. The usual range is between 0.5 and 2.0 times Earth's size. Extrapolating this statistic, if there are about one trillion planets in our galaxy, some thirty billion (or 3 percent) have a mass and radius similar to those of Earth. Of those, a smaller but substantial number (about one billion) orbit G-type stars like the Sun. That's promising but hardly enough. When the focus is on finding life, having a mass and radius *like* Earth's is a long shot from *being like* Earth. Our planet is much more than a rocky world with a certain mass and radius orbiting a G-type star once a year. In its beauty and mystery, life combines astronomical, geophysical, chemical, and biological properties in unique ways. Astronomical variables lay down the groundwork on which life might be possible. But life is a very complex building erected upon this groundwork, demanding a host of extra ingredients and steps that, as we will see, are hard to come by.

The upshot from exoplanetary astronomy's truly spectacular results is that we haven't yet found an Earth 2.0.[42] And even when we do find an exoplanet with mass and radius similar to those of Earth, orbiting a G-type star like our Sun in about one year, that world will *not* be another Earth. It may share many of the astronomical and even geophysical properties of our planet, but it will not be another Earth. Our world is unique. There is only one Earth in this galaxy and, I venture to say, anywhere in the visible Universe. There is no Earth clone. Life changes everything. But before I argue why in part III, we first need to explore the essential question of how we could detect life on other worlds, should it exist.

CHAPTER FIVE

LIFE ON OTHER WORLDS

> Well, if we don't have a distinctive position or velocity or acceleration, or a separate origin from the other plants and animals, then at least, maybe, we are the smartest beings in the entire universe. And that's our uniqueness.
>
> —Carl Sagan, *The Varieties of Scientific Experience: A Personal View of the Search for God*

If life exists elsewhere in our galaxy, there are three ways we could find it. The first and easiest is for aliens to visit us. The second is for us or our machines to travel to other worlds and find life there. The third, and by far the most realistic, is for us to gather evidence of life on other worlds by observing them from here. Let us briefly examine each of these possibilities, starting with the most far-fetched, aliens visiting us.

THE DEEPEST SILENCE

Since there is no evidence whatsoever of intelligent aliens in our solar system, they'd have to travel here from other stars. For aliens from another stellar system to visit us, they would have to be far more technologically advanced than us. As Arthur C. Clarke once remarked in what became known as his third law, "Any sufficiently advanced technology is indistinguishable from magic."[1] Our fastest spaceships would take around one hundred thousand years to arrive at the nearest star system to our Sun, the Alpha Centauri triple star system at a mere 4.37 light-years away. Even light, the unbeatable speed champion of the Universe, would take four years and four months to make the trip. Traveling at one-tenth of the speed of light, something we could conceivably achieve with solar-sailing technology, the trip would still last more than four decades. Clearly, if aliens can cover interstellar distances, they know many things we don't. Their technology would be like magic to us.

Countless sci-fi scenarios dream up futuristic technologies that aliens could use to travel across the galaxy. The most popular ones involve a version of wormholes, or Einstein-Rosen bridges. Wormholes are literally tunnels across space-time that could, in principle, work as formidable shortcuts. Why circle the lake walking along the lakeshore when you can take a boat right across it? Like ordinary tunnels, wormholes have two openings between a tubelike structure. Unlike ordinary tunnels, they are a fold in space and need very exotic physics to exist and to keep their "mouths" open. The challenge increases when something as large as a spaceship attempts to traverse a wormhole. In *2001: A Space Odyssey*, Clarke imagined an advanced alien civilization that engineered wormholes across the galaxy into a network of tunnels like a subway system. If this network exists, it remains utterly invisible to us.

Here is an oversimplified way to visualize a wormhole. Since we are not good at imagining things in three dimensions, let's think of wormholes in two dimensions, like the surface of a table or a balloon. Imagine a very long sheet of paper. That's your "universe." To go from one point to another that is far away takes time. If you bend the paper into the shape of a big *U*, you could crawl along the surface from one side to another, or if there is a wormhole at your disposal—a tunnel connecting the two sides—you could cut right through. Unfortunately, to keep wormholes from collapsing we need a kind of exotic matter that we have no clue where to find. But perhaps the aliens can. That's their "magic."

This kind of argument can lead to endless and aimless (but fascinating) speculation. Why assume that aliens so far advanced technologically are still bound by the chains of aging bodies? As we see our own technology advancing, and our minds becoming ever more entangled with digital devices, we can envision a kind of transhuman future whereby our mind's essence, what we (loosely) identify with our inner self and memories, becomes immaterial, soul-like, tethered to reality through information alone. In his novel Clarke speculated that aliens would have broken away from carbon-based and robotic machine structures so "that the mind would eventually free itself from matter . . . and if there is anything beyond *that*, its name could only be God."[2] This is where astrotheology begins, as we envision aliens as the techno-version of godlike creatures, with the obvious subtext that one day we are going to get there too. So, not only is their technology magic to us, but their very existence becomes equivalent to a supernatural presence—omniscient, omnipresent, and undetectable by our feeble human senses and machines. Such aliens are indistinguishable from gods inhabiting the heavenly realm, being as elusive as countless deities have been throughout human history. They exist only in that intangible dimension of faith.

What about more tangible, space-faring aliens like the ones from *Star Wars*, *Dune*, or *Star Trek*? Unfortunately (or fortunately if you are a pessimist), if intelligent aliens do exist in our galaxy, they haven't visited yet. If they have, they must be extremely shy and efficient at hiding. No artifact built with alien technology has ever been found. Humans, not aliens, built the pyramids in Egypt and Latin America, as well as Stonehenge and other large-scale ancient monuments. Swiss best-selling author Erich von Däniken's fantastical speculations of cave paintings and ancient artwork as depicting astronauts and spaceships have been completely discredited.[3] They are also racist, given that most of those supposedly incompetent ancestors were natives from non-European parts of the world. As Carl Sagan wrote in 1980, "That writing as careless as von Däniken's, whose principal thesis is that our ancestors were dummies, should be so popular is a sober commentary on the credulousness and despair of our times."[4]

If anything, the credulousness and despair of our society only grew during the intervening forty-plus years. I am writing these lines on the day of the first congressional public hearing on UFOs in decades. The big novelty is that UFOs, or unidentified flying objects, are now called UAPs, or unidentified aerial phenomena. The threat, many lawmakers claim, may not be coming from other planets but from experimental weapons from China or Russia. But not everyone is convinced, even if they should be: terrestrial experimental aircraft are much more plausible visitors to our skies than anything alien. What's truly remarkable is not the strange lights in the sky but the deep silence—the silence that points to our cosmic loneliness.

Despite countless UFO sightings and stories of aliens abducting humans, the fact remains that we have no convincing evidence that aliens have traveled across vast interstellar distances to grace us with their wisdom or, more alarmingly, to threaten us into oblivion. An alien visit or encounter, at least for now, remains in the realm of speculative fiction. We discussed some of these fictional works before, from Lucian in ancient Rome to Kepler in the seventeenth century to H. G. Wells at the turn of the twentieth. As our more recent fascination with countless sci-fi movies and novels shows, the imagined alien, the "other from outer space," has always been a mirror to humanity. "They" will do to us what we have done to ourselves. They are the tribes that invade the lands of other tribes to pillage, to kill, to rape, and to enslave. They are Western colonizers who spread across the Americas, Africa, and Southeast Asia in search of material goods and economic expansion, with complete disregard for the cultural values and freedom of the Indigenous peoples and for the natural environment. They are expansionist empires with a culturally constructed ideal of superiority that crush smaller neighbors along their way. They are the horrors of the Holocaust and countless genocides across history.

"Us and them" has always been about us. The fear of the other is a deep-seated fear of ourselves, of what we humans are capable of doing to our fellow humans. We project onto the eerie silence of the stars the anxiety of our cosmic loneliness, the fear that ultimately we alone must decide whether we will overcome our destructive greed or instead let it drive us to our collective end.

This is the moral impasse we must face, and which must end if we are to preserve our project of civilization.

FLYING TO THE MOON (AND BEYOND)

A multitude of worlds are out there, waiting for us among the distant stars. But traveling interstellar distances is at present impossible. Not that there is any law of Nature that prohibits long-distance space traveling; the barriers are both physiological and technological. Our propulsion systems are not fast enough to send us to the stars. Our bodies are too frail. In space, we lose bone and muscle mass, and our bodies are susceptible to radiation. We also suffer under prolonged isolation and solitude.

We evolved to be on this planet, under very specific conditions. When we leave Earth, we take some of it with us, what we need to survive: air rich in oxygen, a balanced temperature, our foods and medical supplies. But this small-scale hauling of an Earthlike environment is very limited and costly. What we can do, and have done spectacularly well, is send our probes to the worlds we can reach with our current technology—the worlds of our solar system. The past few decades will be known in history as the solar system exploration era. We have landed rovers on Mars and have sent probes to all seven planets of our solar system, and also Pluto, now demoted to a dwarf planet. Voyager 1 and 2, both launched in 1977, are now exploring interstellar space beyond our solar system. We have mapped, measured, sampled, photographed, and orbited countless alien worlds in our solar neighborhood. The Martian rovers are actively scouring the planet's surface, and now the subsurface, for hints of life. Flybys of Jupiter and Saturn moons have revealed oceans underneath thick ice crusts, active volcanoes and geysers spewing water-vapor plumes mixed with organic compounds, lakes and rivers of flowing liquid methane and other organics. What we have found orbiting our Sun is but a preview of the staggering variety of worlds spread across the vastness of our galaxy. And, at the same time, the harshness of those alien environments inspires a deep appreciation for the uniqueness of our home planet. Space is horribly hostile to us humans.

Between July 1969 and December 1972, NASA landed six crewed missions on the Moon. A total of twelve humans left footprints on our barren satellite world. Since then, no human has stepped on another world. Sending machines to explore worlds beyond the Moon is safer and economically more viable. Still, we should expect humans to fly to Mars within the next few decades, continuing our slow spread across the solar system. But it's hard to contemplate human visitation beyond Mars; the costs are enormous, in terms of both finances and health.

In Clarke's 1968 novel *2001: A Space Odyssey*, 2001 was the year humans reached Saturn (Jupiter in Stanley Kubrick's movie).[5] This was very optimistic, given that we haven't even landed on Mars yet. But when I first watched the movie in 1969, my ten-year-old imagination exploded with possibilities. How far could we venture into outer space? What would we find? How could I be part of this quest? Back then, the year 2001 felt very far away, an almost dreamy time in the future. Futuristic science looked like magic, but a magic that *we* made happen, not aliens. I realized then that I wanted to be this kind of magician, the kind that flirts with the unknown to advance human knowledge, transforming imagination into reality. We may not have landed on Mars or traveled beyond to Jupiter or Saturn, but we have sent probes to map the confines of our solar system, discovering worlds of wonder and mystery in our search for answers. So far, we have found no traces of life, past or present. Despite our longing for cosmic company, our solar system seems barren.

LISTENING AND LOOKING FOR LIFE ELSEWHERE

If aliens haven't visited Earth and life in our solar system seems to be confined to this planet, we need to search farther out. Since any interstellar travel technology is a dream far in our future, what we can do now is look and listen. Listening for aliens is the central goal of the Search for Extraterrestrial Intelligence (SETI) Institute, a five-decade-long ongoing effort to detect and decode radio signals emitted by alien technological civilizations in our galactic neighborhood. Despite a few hopeful signals, we haven't succeeded in detecting anything promising. There are many possible reasons for this, which we will discuss soon. For now, we can state that the deep silence persists, despite the remarkable dedication of hundreds of SETI scientists over the years.

If listening is not working, we can look for other signs of life elsewhere. Some SETI projects have used powerful telescopes to find signs of planetary or even stellar-scale engineering in distant star systems. Although this is a very tantalizing possibility, the more realistic and immediate way to find signs of life on distant worlds is to search for *biosignatures*, that is, signs that biological processes may leave in the atmospheres of exoplanets. This is where most astrobiologists would place their bets for the highest chances of success. The widespread presence of life on a planet changes the planet and its atmospheric composition. As my past graduate student Sara Imari Walker, now a professor at Arizona State University, once quipped, "Life doesn't happen on a planet; it happens *to* a planet." This is most definitely true on Earth, and we have every reason to believe that it will be true for other life-bearing worlds where life takes hold and spreads to become an active, planetary-scale biosphere.

We now know that the vast majority of stars have orbiting planets, and we know enough of them to estimate the distribution of alien worlds in our galaxy, grouping them as Neptunes (gas giants about the size of Neptune), hot Jupiters, super Earths, and terrestrials (a rocky world with a radius between 0.5 and 2.0 times that of Earth). Given that the formation of stars follows the same basic laws of physics across the Universe, we should expect this to be true everywhere, in our galaxy and others. Planets are a bit like snowflakes: they all share certain basic properties, but no two are alike. The question, then, is whether a subset of these alien worlds has traits similar to Earth's. In other words, how common or rare is our home planet?

An identical clone to our planet would be impossible to find, given that each world has its own very specific history, including different distributions of chemical elements, different distance to its host star, different neighboring planets, different number and sizes of moons, and different impact history. A rocky planet with mass and radius nearly identical to Earth's that is orbiting a G-type star like our Sun once a year is *not* another Earth. Sharing many astronomical properties is just the basic set of requirements—

useful for identifying worlds that deserve further scrutiny as potential harbors for life as we know it. But, as we have hinted and will explore in detail, much more is needed for life to exist and endure on a planet or moon. The story of life and of Earth must be told together for us to understand why Earth is such a special world.

The transit method, discussed in the last chapter as a way to find exoplanets, also offers the best chance we have to pinpoint planets that may harbor life. When a planet passes in front of its host star, some of the starlight gets absorbed by the planet's atmosphere. This happens because each chemical element and molecule absorbs and emits light at specific wavelengths, known as *spectral lines*. We call the collection of spectral lines, the *spectral signature* of the chemical. Just like people have unique individual fingerprints, calcium has its own spectral lines, as do hydrogen, methane, oxygen, ammonia, and so forth.[6] Since each planetary atmosphere has a unique combination of chemicals, it will thus have a unique spectral signature. We call these spectral signatures the *absorption spectrum* of the atmosphere, since the chemicals in the atmosphere are absorbing some of the starlight. Astronomers collect the absorption spectrum of the planet and then identify wavelengths specific to each chemical element in the atmosphere. By reading the different lines in the spectrum, they then know the composition of the planet's atmosphere. Does it contain water? Carbon dioxide? Methane? It's astronomy detective work, incredibly important to the search for life on other worlds.

Life, when abundant, imprints itself on a planet's atmosphere. If we know the chemicals related to life, we can then search for those in the planet's absorption spectrum. These are the biosignatures. Earth as seen from far away would be our guide. An alien studying our atmosphere would identify water, carbon dioxide, oxygen, methane, and ozone, among other chemicals. This combination is a signature that life is present and active, interacting with the planet's atmosphere and leaving its mark. Just finding water or carbon dioxide is important but not enough. A planet with life is a dynamic engine, where biological and geological processes combine and feed back on each other. When life takes hold of a planet, it cannot be separated from it. Planet and life form a single whole. A living planet like Earth heaves and breathes as plants and animal life breathe through the daily and seasonal cycles. Life changes the planet. The planet changes life. The history of life on a planet and the planet's life history are inseparable, woven together through the ages.

PART III

THE UNIVERSE AWAKENS

CHAPTER SIX

THE MYSTERY OF LIFE

> But if (and oh what a big if) we could conceive in some warm little pond with all sorts of ammonia and phosphoric salts,—light, heat, electricity etc., present, that a protein compound was chemically formed, ready to undergo still more complex changes, at the present such matter would be instantly devoured, or absorbed, which would not have been the case before living creatures were formed.
>
> —Charles Darwin, letter to Joseph Dalton Hooker, February 1, 1871

A PERSISTENT PUZZLE

Life, ubiquitous as it is, remains a profound scientific mystery. Perhaps surprisingly to many, at present scientists share no agreed-upon definition of life or, for that matter, an understanding, even at a primitive level, of how life originated on Earth. To rephrase, we don't know what life is or how it began. These two codependent questions relate to life and its nature. By "codependent," I mean that it's hard to imagine what life is without also understanding how it came about. We know of life only on one world—ours—and our thinking is necessarily biased toward what we know. When scientists talk about finding life on other worlds, they are usually referring to life as we know it. The go-to operational definition NASA adopts states that life is a self-sustaining chemical reaction network capable of reproduction that undergoes evolution through Darwinian natural selection. So life metabolizes energy, reproduces, and undergoes changes. But knowing so much about our form of life doesn't mean we know what it took for life to emerge here. We can't travel back four billion years to primal Earth to learn how a soup of inorganic compounds became, after a few steps, a soup of organic compounds—the stuff of living things. And then, more enigmatic still, this soup of organic compounds ended up trapped inside a membrane and found a way to eat and to self-reproduce. Somehow, somewhere on primal Earth, inanimate matter became living matter and the first single-celled organisms emerged.

Each of these steps and the many that followed in the evolution of life on Earth were hugely complex and nonpredictive. Currently we have at most a very fragmentary understanding of what could have happened here billions of years ago. Worse, some of the details are unknowable, lost in the fog of time. Finding evidence of very early life is not the same as finding evidence of *first* life. How could we ever know if some piece of evidence we find of early life can be tagged as the very first life on Earth? Even if someone is able to synthesize life from nonlife in the laboratory, how can we know that this was the path life took to emerge here more than 3.5 billion years ago? Although scientists don't like to admit this, when it comes to the origin of life, we have to resign ourselves to a story with the beginning missing. The origin of life on Earth is unknowable.

But unknowable questions are not a deterrent to knowing

more. They are an inspiration. Even if we can't get to a final understanding of what happened on our planet more than 3.5 billion years ago, we can learn a tremendous amount while trying. Astronomy has revealed a mind-boggling number of worlds out there. It is thus natural to expect that life should have emerged on many of them, or may still do so. Biology reframes these expectations, turning the origin and evolution of life on other worlds into a one-of-a-kind experiment, with its own unique outcomes, if any. When considering life elsewhere, we need a post-Copernican worldview in which astronomy is woven into biology. The existence of many worlds, even many Earthlike worlds, does not mean the existence of many living worlds.

The epigraph that opens this chapter, from a letter Darwin wrote to a friend, summarizes his thoughts on the matter: some sort of watery chemical soup rich in nitrogen and phosphorus compounds, bathed in sunlight, warmth, and electricity, got the first steps toward life going—the formation of amino acids that link up to form simple proteins. Following Darwin's speculations, we could say that these initial ingredients—a prebiotic soup, partially isolated from the outside environment by some kind of baglike membrane, possibly a droplet of fat (a lipid boundary)—would then evolve into more complex chemical compounds, eventually becoming a chemical reaction network capable of self-reproduction: life! To Darwin, and to many scientists still today, *abiogenesis*, the transition from nonliving matter to living matter, happened in Earth's primitive environment, possibly triggered by some kind of electric stimulation, often conjectured to be lightning during volcanic activity. One can't avoid but think of Victor Frankenstein and his macabre experiment. Others, most notably

Svante Arrhenius and more recently Iosif Shklovskii and Carl Sagan, as well as Francis Crick and Leslie Orgel, have defended the idea that life was seeded on Earth from outer space, a process known as *panspermia*.[1]

Even though the hypothesis of panspermia is fascinating, it pushes the origin of life to another world or to the machinations of an alien intelligence, not helping us understand any better how these cosmic life seeds come about in the first place. If we somehow confirmed that life on Earth came from another world, intentionally or not, we still wouldn't know how it originated there. In fact, panspermia as an explanation for the origin of life on Earth is the biological equivalent of the "turtles all the way down" answer to the origin of the Universe.[2] If one postulates that the Universe came from an initial quantum state, as modern models of quantum cosmology do, we can always ask "And where did that particular quantum state come from? What set its specific properties?" Simple causal logic fails when we consider the start of an unknowable chain of causation, the First Cause, the cause that can't be caused. (Otherwise, it would need a previous cause, and another one, and another one . . .) Like the origin of the Universe, the origin of life also suffers from a First Cause problem, given that at this point we don't know how to frame the enigmatic passage from nonlife to life in causal terms. There are cosmogonical (referring to the origin of the Cosmos) and biogenic (referring to the origin of life) First Cause blind spots that frame the boundaries of scientific explanations.[3]

Research on the origin of life is now a robust and thriving scientific discipline. Some consider it a branch of astrobiology, whereas others take it to be an independent field of inquiry rooted in molecular biochemistry and cell biology. The general attitude of scientists is pragmatic, a position that could be called "shut up and experiment!" as opposed to delving into the murky issues of defining or interpreting the nature of life.[4] In the biological sciences, the laboratory is the crucible of controllable knowledge about life. Quite reasonably, scientists naturally tend to focus on questions they can explore in the laboratory. While a minority hopes that experimental results will illuminate more fundamental conceptual challenges, the majority adopts a more pragmatic approach and ignores such fundamentals altogether as useless nuisances, claiming that philosophical musings won't advance scientific knowledge. It is thus not surprising that fundamental conceptual questions remain unanswered or unaddressed.

As an illustration, consider the so-called RNA world hypothesis, based on the assumption that genetics precedes metabolism in the unfolding of life and that the RNA molecule takes the driver's seat before DNA.[5] Although this hypothesis presents a very compelling possibility along the evolutionary line from primitive to more advanced life, the question is whether RNA experiments can really tell us something fundamental about life's origin or how life first achieved its reproductive capabilities. After all, RNA molecules are extremely complex, composed of billions of atoms and capable of both storing genetic information and catalyzing chemical reactions. It is hard to imagine that life's evolutionary roots don't reach farther back to much simpler early reproductive systems. It also seems reasonable that those simpler reproductive systems had to first metabolize energy before they could reproduce. Any life must eat before it can breed or multiply.[6]

Even if RNA world–type experiments are clearly essential

to elucidate some of the conundrums related to molecular evolutionary mechanisms,[7] it's hard to see how they can take us to the earlier stages of emerging life. By way of comparison, reverse engineering a SpaceX rocket ship won't teach us much about the early history of aviation, with its balloons and dirigibles.

Geology brings on another complication to the RNA world hypothesis, given that there is scant and at most indirect extant evidence of primitive metabolic activity imprinted in rocks from the time life is (currently) believed to have taken root on Earth, somewhere between 3.8 and 3.5 billion years ago. From an origins-of-life perspective, the RNA world scenario is attractive mostly because scientists can experiment with these systems, and not so much because their complex adaptive molecular dynamics can be reverse-engineered back to their distant relatives during life's first hesitant steps, when growing numbers of carbon atoms began to chain together to form longer molecules (polymers). The situation is a bit like a fellow looking for his car keys at night in a big parking lot. He will search near streetlights because that's where he can see best, not because he knows that's where he dropped them. He may even find interesting objects near the streetlights; but given the large surface area of the search space, odds are he dropped his keys far away from the lights, somewhere in the darkness.

WHY IS LIFE SO DIFFICULT TO UNDERSTAND?

Despite spectacular advances in biochemistry and genetics research, it is hard to see how the question "How did life emerge on

Earth?" could be answered. This doesn't mean that we are saying that the emergence of life is some kind of supernatural phenomenon. Not at all. Life is a very natural phenomenon. The problem scientists face is one of information retrieval, that is, of trying to reconstruct an era lost in the distant past with an appalling scarcity of clues. Given the irretrievability of information about the specific environmental conditions and biochemical pathways that led to first life on Earth about 3.5 billion years ago (or before? We aren't sure), how could we hope to falsify proposed mechanisms? As we mentioned above, how could we be sure that detectable life-related signatures from Earth's infancy are illustrative of first life and not later life? The conventional strategy, a laboratory-based, bottom-up approach that would coax a biochemical reaction network to transition from nonlife to life, even if successful, could not be proved to be equivalent, much less identical, to what happened during Earth's distant past. In other words, unless it could be formally proved that there are only very few possible biochemical pathways from nonlife to life, or even better, only one, the option of creating life in the laboratory—no doubt a spectacular feat if ever achieved—wouldn't tell us anything about how life first emerged on Earth.[8]

Moving away from Earth to consider identifying life elsewhere creates new challenges. Unless one could prove that life follows the same laws across the Universe (or just be lucky enough to see signals of it), there is no guarantee that life on other worlds would be relatable to life on Earth. Even if we could somehow map the biochemical pathway or pathways that led to life on Earth, or "life as we know it," we wouldn't necessarily learn anything universal about the nature of life on other worlds. The certainty that physical scientists have that the laws of physics and chemistry are the same across the Universe cannot be duplicated in the biological sciences. Biology is nonmechanistic. Evolution by natural selection depends on the randomness of genetic mutations, on unpredictable complex fluctuations of environmental conditions, on nonlinear feedback loops that couple diverse geophysical phenomena to life itself, on cosmic and global cataclysms that reset the evolutionary stage, and so on.

Many scientists and philosophers argue that insisting on applying what we know of life on Earth to other worlds will hinder rather than help our success in identifying and studying it.[9] These thinkers contend that definitions are constraining, since they box meanings inside a volume that may be smaller than what is needed to tackle such a vast problem. However, if we don't have a definition or an understanding of what life is or could be, how can we hope to re-create it in the laboratory or be sure to identify it on another world, where life could possibly follow very different rules? NASA and other research agencies are beginning to recognize this issue. In 2019, NASA unveiled the Laboratory for Agnostic Biosignatures, aimed at supporting scientists who are thinking outside the box to develop strategies for identifying signs of unusual life-related activity.[10] The essential shift here is to let go of the "what life is" conundrum to focus on the more pragmatic "what life does."[11] For example, novel signatures of alien living systems could manifest themselves in terms of chemistry with levels of complexity that go well beyond what known inorganic or simple "nonliving" organic chemistry can achieve; or worlds may have peculiar atmospheric chemistries that indicate metabolic activity of some kind.

When it comes to types of life, we are restricted to a statistical sample of one—life on Earth. Even if we begin to find ways to differentiate between terrestrial and alien life-forms, this differentiation hinges on "life as we know it" versus "life as we don't know it." The complication, though, is that we can't even start from "life as we know it" with certainty, given that there is so much we don't know about it. Our imagination is limited by our compromised objectivity, given that our existence on this planet is contingent on how life coevolved with the planet. Science and logical reasoning are our best tools to avoid cognitive bias, but they are never foolproof.

It would indeed be wonderful if we could affirm, with the confidence of physicists and chemists, that the laws of biology are the same across the Universe. Physicists and chemists can do this because they have information about physical and chemical processes occurring across space and time that confirm the universality of physical laws: the conservation of energy, the fundamental forces between elementary particles of matter and the electrical forces that form atoms and molecules, the ubiquitous power of gravity across vast distances, the ninety-four naturally occurring chemical elements and their formation in stars and through radioactive decay.

Still, evolution by natural selection is an extremely powerful idea, and it's hard to imagine that life elsewhere could exist without it. After all, any form of life needs resources in a limited environment that is subject to change. Life eats. And because it eats, it needs to find food. So, living creatures forage. If they can't move in space, they grow deep roots and stretch upward in search of sunlight. But even with that, the vast combinatorial space that frames viable life-forming chemistries and life-supporting environments—added to the also vast conceptual barriers we have to understanding the many steps from nonlife to microbial life and, from there, to complex life—precludes hopeful generalizations. As physicist Philip Anderson once remarked, referring to emergent complex systems in Nature, "More is different." New laws emerge as matter organizes at increasing levels of complexity that cannot be traced down to simpler ones in typical reductionistic fashion. To say that one can successfully describe the workings of a cell starting from quarks and electrons is an epistemologically incorrect fantasy, even if, ultimately, cells are made of quarks and electrons. To quote Anderson, "The ability to reduce everything to simple fundamental laws does not imply the ability to start from those laws and reconstruct the universe."[12]

Life's evolution is an expression of a complex compromise between bottom-up forces and top-down environmental causation. What in physical systems are called boundary conditions, the external constraints that determine the ultimate dynamics of a system composed of many interacting parts (these constraints could be the shape of a container, the viscous properties of a diffusive medium, external factors like changing temperature or pressure that force the system into different states, and many others), in living systems becomes a manifold challenge with vast unpredictability. Any workable model must drastically reduce the number of acting variables and thus have limited applicability.

The great biologist Ernst Mayr listed some of the distinctive characteristics of the biological sciences that necessitate a different way of conceptualizing life: "The rejection of strict determinism and of reliance on universal laws, the acceptance of merely probabilistic prediction and of historical narratives, the acknowledgement of the important role of concepts in theory formation, the recognition of the population concept and of the role of unique individuals."[13] Theoretical biologist Stuart Kauffman eloquently defends this view when writing about the evolving biosphere: "a becoming biosphere, a rich and almost unfathomable becoming . . . in ways we cannot prestate, yet is somehow coherent. Despite bursts of extinction events and the fact that 99 percent of all species are gone, the biosphere flowers on. And on and on and on, ever-becoming beyond what we can say ahead of time."[14] Given the first single-celled organisms, no one would have predicted dinosaurs. There are profound differences between nonliving and living matter. Determining these differences is no small task.

CAN WE TELL LIVING FROM NONLIVING?

As an illustration of such definitional difficulties, consider three very different physical systems: fires, hurricanes, and stars. Although all three have general thermodynamical properties that we use to describe life, grouped together as *nonequilibrium dissipative structures*, we know that fires, hurricanes, and stars are not alive. As we figure out the differences between these systems and living beings, we learn more about what life *does*, moving away from the harder question of what life *is*, while acknowledging that these two aspects of life are woven into an irreducible tangled whole.

Let's first consider fires. To sustain themselves, fires spread and feed on their environment. They consume oxygen to keep on burning and are thus open thermodynamical systems, as are living creatures. Given the right conditions, fires multiply, often with devastating consequences. But I know and you know that fires are not alive. We wouldn't consider the spreading of a fire as a form of reproduction. We wouldn't call oxygen combustion a form of metabolic process. Why is that? For starters, fires don't have a history. They don't have a gene storage mechanism they use to transmit their characteristics as they spread. They also don't have survival strategies or repairing mechanisms. If a fire is burning down a ravine toward a creek, it will keep burning until it stops by the water and eventually dies out. It doesn't forage intentionally for more fuel or strategize in any way to continue burning.

What about hurricanes? Like fires, they are persistent far-from-equilibrium complex systems (as are living creatures) that need the right environmental support to exist and to maintain themselves. They "move" and are tightly coupled to local humidity, pressure, and temperature conditions. If favorable atmospheric conditions hold, they maintain their basic shape. Jupiter's Great Red Spot is a giant anticyclonic storm that has endured for at least four hundred years. But as with fires, we wouldn't equate these properties of hurricanes with being alive.

Stars are similar. They are self-sustaining, in the sense that they convert gravitational potential energy (they are slowly imploding) into the enormously high pressures and temperatures at their cores that promote the nuclear fusion reactions they need to sustain themselves. One could even say that they are self-cannibalizing entities that "eat" their own entrails to survive. Stars exist because of a constant tug-of-war between gravity trying to squeeze them inward and nuclear fusion trying to blow them apart. Amazingly, this seemingly unstable and very dramatic situation can render stars stable for billions of years. Our Sun, a modest-size star a little less than five billion years of age, is at roughly the middle of its "life cycle" (note the nomenclature). Stars form in regions rich in gases and chemical elements called stellar nurseries (again). When a star exhausts the fuel at its core, it dies (and again) in a huge explosion, creating shockwaves that propagate and spread its material across interstellar space. When these shockwaves from the dying star collide with gas clouds, they can trigger the formation of new stars. In a loose sense, then, stars are reproducing, even sharing some of their original matter with the nascent ones. Still, you know and I know that stars are not alive.

There is poetry in the cycle of the life and death of stars, made meaningful by our own experiences of life. We are so imbued with life that we tend to see it everywhere. It is, perhaps, a manifestation of Francis Bacon's "idols of the tribe," one of the four barriers to the truth he called "idols and false notions."[15] We tend to overgeneralize and jump to conclusions too quickly, often ignoring evidence against our views. The result is that we are easily deceived, because we want so badly to believe. This is where science comes in, as a powerful antidote against magical thinking, albeit fallible. Still, we can use scientific reasoning to distinguish between the living and the nonliving, even if sometimes the boundaries between the two are subtle.

An essential difference is that living systems have an unpredictable aspect during reproduction, a random variability that is absent in nonliving systems. Indeed, for physical systems, if we repeat initial conditions with very high precision, a fire would always burn the same way, a hurricane would spin the same way, and a star would evolve the same way, even if small details would vary. It is as if nonliving systems have an information content that is nearly frozen—a repeatable history from inception to end—while living systems have an information content that is fluid—an unpredictable history from inception to end. Fires and hurricanes don't evolve from ancestors.

Living creatures take advantage of available energy to remain stable in the face of changing conditions, a property known as *homeostasis*. They are called open thermodynamical systems because they absorb energy from the environment and get rid of what they don't need. For example, if it's hot, we use nutrients and water to sweat to keep the body cool. Nonliving dissipative structures, such as hurricanes, tornadoes, convection cells, and turbulent flow, are also open thermodynamical systems that take advantage of available energy to reach a steady state, maintaining their overall spatial structure while conditions are supportive (picture a hurricane moving along the Caribbean Sea toward Florida). The essential difference is the passivity of nonliving dissipative structures when contrasted with the active behavior of living systems. Life strategizes to find nutrients even at the bacterial level (chemotaxis), sensing the best path forward through a yet unknown interplay of bottom-up and top-down causation. We use words like "volition," "urge," "autonomy," and "control" to describe living systems and even biospheres but wouldn't use such words to characterize fires, hurricanes, or stars. Still, the puzzle of how life emerges from nonlife remains as mysterious as ever. How does an agglomeration of inanimate matter, beyond an unknown level of biochemical complexity, become a living creature? We don't yet know how to think about the transition from nonliving to living, or how a bundle of inanimate chemicals turns into an entity with a sense of purpose.

THE VIRTUOUS CREATIVE CIRCLE OF LIFE

Life is a process that captures energy and metabolites to sustain itself and reproduce. Living creatures are thus a paradox, given that they are at once a unit that sets itself apart from the background where they exist and are actively dependent on and inseparable from that background. As Chilean visionary biologist Francisco Varela wrote: "A cell stands out of a molecular soup by defining and specifying boundaries that set it apart from what it is not. However, this specification of boundaries is done through molecular productions made possible through the boundaries themselves." To illustrate his point, Varela uses M. C. Escher's famous drawing of two hands popping out of a page to draw themselves: "The cell draws itself out of a homogenous background," thus both standing alone as itself while emerging and being part of the background from whence it emerges. The boundaries that specify the living cell—essential as they are—are fuzzy, merging "producer and product, beginning and end, [and] input and output."[16]

Where then do we draw the boundary between the living and the nonliving, given that the two are inextricably connected? The air that we breathe, the warmth that sustains us, the food that we eat, the complex bacterial biome in our guts—they are also us and we are also them. Who we are, our being, extends beyond the boundary of our physical bodies. The *process* of life necessitates a connection with the outside, blurring the boundaries between the living and the nonliving in truly puzzling ways. While we are intuitively drawn to distinguish the living from the nonliving, the living "I" from the nonliving "others" that surround the I, the distinction is both obvious and unclear. You know you are you and not the air you breathe or the food you eat. But you are also enmeshed with both and couldn't be yourself without them being within you and outside of you.

Varela called this closure a strange loop, a "virtuous and creative circle." Strange, because it clashes directly with science's presumed objectivity, which is based on the assumption of a clear separation between observer and observed. Indeed, biology is frequently contrasted with quantum physics precisely for being the science where this separation is most clear. We can use a microscope to see the bacteria swimming in the molecular soup in a petri dish. This sounds very clear-cut and objective; however, the molecular soup in the petri dish affects the bacteria swimming in it, and the bacteria, in turn, affect the molecular soup. The observer makes choices, and these choices affect the petri dish and the bacteria swimming in it. There is an *entanglement of autonomies* here, from the bacteria and the molecular soup in the petri dish to the scientist in the lab. From the observer's perspective, everything that happens during observation relies on the observer's experience of being in the lab, recording measurements while breathing the surrounding air and metabolizing lunch. This entanglement of autonomies doesn't end here. This blurring of boundaries has critical consequences for how we define the biosphere—the totality of life at a planetary scale—and our relation to it. The strange loops link into the chain of life, making no sense individually. Take one link from the chain, and that life ceases to be possible, its demise affecting other loops. To be viable, life blurs its own boundaries.

Where, then, does a living creature begin and end? A forest is a connected wholeness of trees, fungi, bacteria, animals, insects, birds, each with specific functions, all interdependent. While each individual enacts the hard work of staying alive, foraging, eating, killing, escaping, breathing, nesting, reproducing, reaching deeper into the soil or upward toward the sunlight, there is a unity of purpose in the variety of actions, a shared urge to be, to remain alive. Eliminate one species or change the environment beyond a certain critical level and the forest's integrity is compromised. Life is a collective.

HOW COMMON IS LIFE IN THE UNIVERSE?

Ask most astronomers and physical scientists about life in the Universe, and the answer tends to go like this: Well, consider our galaxy, the Milky Way. It contains about one hundred billion stars, and now we know that most of them have planets. Rounding the numbers up, we are talking a trillion planets or more. Add moons as potential life-harboring worlds, and we easily reach trillions. Each world is different. Each world has an incredibly rich history, contingent on its parent star, the available chemicals in the region, and the world's formation and evolution details. It is a lot of real estate. If we want to narrow down the sample, the current estimates are that about 7 percent of stars in the galaxy are "G dwarfs" like our Sun. That's about seven billion other Sunlike stars out there. Add to this the fact that recent observations indicate that each Sunlike star harbors between 0.4 and 0.9 rocky planets in its habitable zone, and we arrive at the staggering number of roughly three billion or more rocky planets with the potential to harbor life in our galaxy alone.[17]

According to this "large-numbers" astronomy-based argument, life should be ubiquitous throughout the galaxy. Note that the argument says nothing about what *kind* of life we should expect, simple versus complex, unicellular versus multicellular, intelligent or not. It can't, of course, given that it's estimating only the possibility that a world can harbor life on the basis of its rocky composition and potential for having liquid water on its surface.

To the estimates for the huge number of rocky, and potentially water-bearing, worlds, physical scientists (especially cosmologists) often add the mediocrity principle,[18] an heir to Copernicanism when applied to astronomy, as we have seen in chapter 2: there is nothing special about our galaxy, our Sun, Earth, or the evolution of biological complexity observed here, including the diversity of species and intelligence. It follows that life, and even intelligent life, should be common on Earthlike planets throughout the Universe. Under the mediocrity principle, we are the rule, the mediocre majority, and not the interesting and relevant exception. Given that we now have an estimate for the number of such worlds, physical scientists typically conclude that we should expect to find life in millions upon millions of other worlds in our galaxy alone. The mediocrity principle is a sad example of inductive thinking gone crazy.

Fortunately, not everyone, even within the physical sciences, agrees. In 2000, geologist and evolutionary biologist Peter Ward and astrobiologist Donald Brownlee published *Rare Earth: Why Complex Life Is Uncommon in the Universe*.[19] Ward and Brownlee correctly argued that being a rocky planet with surface water is not nearly enough to determine the existence of life on a planet, much less complex life, at least complex life as we know it. (This restriction is important. We don't know how to qualify the possible existence of life as we don't know it, as we mentioned above.)

Ward and Brownlee distinguish the requirements for simple (microbial) and complex (multicellular) life, tying them to the properties the planet may have. Earth, as the only example we have at hand, has several geophysical properties that conspire to provide the long-term stability that life needs to *possibly* evolve from simple to complex through the process of natural selection.[20] Long-term stability, here, doesn't mean the planet doesn't change over the eons (Earth has changed a lot in its history), but that the changes, even when extreme, still allow for life in some niches to survive, mutate, and readapt to the inevitable and often dramatic random environmental changes that unfold over the course of billions of years. Massive volcanic eruptions, dramatic tectonic plate drift and continent formation, and devastating asteroid and cometary collisions are among the causes of the known five global extinction events that occurred during the past 440 million years. We are now witnessing a sixth wave of extinction, known as the Holocene extinction or, to many people, the Anthropocene extinction, given the correlation between the fast dying rate of animal and plant species during the past ten thousand years and the aggressive presence of humans on the planet.[21]

Earth's geophysical properties that act to protect life include, among others, plate tectonics, a large single moon, and a magnetic field strong enough to serve as a shield against life-destroying radiation from the Sun and outer space. Those properties make Earth a much rarer world among rocky planets. Even so, life here came very close to total sterilization. This adds to Ward and Brownlee's argument that when it comes to life elsewhere in the Cosmos, simple life will be much more probable than complex life, which is more fragile to environmental change.

In 2015, Peter Ward joined forces with Caltech geophysicist Joe Kirschvink to publish an updated version of *Rare Earth* called *A New History of Life: The Radical New Discoveries About the Origins and Evolution of Life on Earth*.[22] Ward and Kirschvink outline the steps leading to the transition from no-life to a living protocell:

1. The synthesis and accumulation of small organic molecules, including amino acids and nucleotides. Phosphates are also important, given that they are the backbone of RNA and DNA.
2. The joining of such ingredients into larger molecules, such as proteins and nucleic acids.
3. The aggregation of proteins and nucleic acids within fatty droplets to form the first protocells.
4. The ability to replicate the large complex molecules to establish heredity.

While step 1 can be accomplished in the laboratory, the artificial synthesis of RNA and DNA is much more prohibitive. These are extremely complex molecules that break down when heated, suggesting that they were first made in cold or moderately warm environments. It is highly plausible that life experimented with many primitive molecular replicators before hitting on RNA. What those earlier replicators were remains a mystery.

Once a protocell is made, biology is in full action. From here, the path that starts with prokaryotic cells and leads to complex, multiorgan creatures is fraught with enormous challenges and unknowns. This is where we find, broadly speaking, a departure between physical and biological scientists concerning the ubiquity of life in the Cosmos.

In a previous book, I listed the nine steps from no-life to intelligent life as follows:[23]

(1) Inorganic chemistry → (2) Simple organic chemistry → (3) Biochemistry → (4) First life (protocells) → (5) Prokaryotic cells → (6) Eukaryotic cells → (7) Multicellular life → (8) Complex multicellular life → (9) Intelligent life.[24]

The first four steps of this list align well with those of Ward and Kirschvink described above. However, to arrive at intelligent life there are five extra steps (steps 5 through 9), all extremely complex and quite possibly highly unlikely. (We can't be precise in terms of probabilities, given that we know life only here on Earth.) Here they are in more detail, picking up from step 5:

1. **Protocells to Prokaryotic Cells:** The transition steps from complex proteins and nucleic acids to primitive protocells and then on to the first prokaryotic cells are unknown. Presumably, a protective membrane

made of fatty molecules surrounded the reacting chemicals, isolating them from the outside environment. (Fat droplets are hydrophobic, that is, they keep water away.) With increasing efficiency, the membrane allowed energy and nutrients to come in and waste to get out. Meanwhile, the genetic material inside the primitive cells replicated, leading to fast diversification. This was the world of protozoa, with natural selection driving protocells toward increasing metabolic and reproductive efficiency through trial and error. Life emerged without a plan.

2. **Prokaryotic to Eukaryotic Cells:** We have little understanding of the next step in life's complexity—the emergence of eukaryotic cells from prokaryotic cells, although we do know that the transition took close to two billion years. The most accepted view, suggested by biologist Lynn Margulis, is that eukaryotes developed from symbiotic alliances between prokaryotes. For example, mitochondria, the modern cell's little engine, are believed to have been a separate organism in the distant past that was either eaten or absorbed by another cell.[25]

3. **Unicellular to Multicellular Life:** Then comes another crucial step, the transition, roughly three billion years after life's first known traces, from unicellular to multicellular organisms. As with the transition from prokaryotes to eukaryotes, multicellular organisms possibly also evolved through symbiotic trial-and-error processes, as different kinds of unicellular organ-

isms linked to each other (or ingested each other) and became pluralistic in form and function. However, it's hard to understand how the different types of DNA from diverse organisms became incorporated into a single genome. As an alternative explanation, the Colonial Theory proposes that unicellular creatures grouped in colonies that slowly evolved into multicellular animals. Although the debate is still on, the Colonial Theory continues to gain adherents.

4. **Multicellular to Complex Multicellular Life:** Many scientists propose that environmental changes played a major role in accelerating the diversity of complex multicellular organisms that climaxed during the so-called Cambrian explosion, about 530 million years ago. Chief among them were the rapid increase in oxygen availability in the atmosphere and the advent of continent formation and plate tectonics, with the resulting remixing of surface and ocean chemistry. Tectonics work as a global thermostat, recycling chemicals that help regulate the levels of carbon dioxide in the atmosphere and keep the global temperature stable. Without it, surface water would not have remained liquid for billions of years, and life, especially complex life, would have faced insurmountable obstacles.

5. **Complex Multicellular to Intelligent Life:** After about five hundred million years of evolving multicellular organisms, including many severe mass extinctions and climate changes, the first members of the genus *Homo* appeared in Africa some four million

years ago. Intelligence as we know it is fewer than one million years old. It's been around for less than 0.02 percent of Earth's history.

Contrasting the viewpoints of the physical and biological sciences, it is hard (if not plain wrong) to justify naive optimism that life, and in particular complex intelligent life, is ubiquitous in the Cosmos. There is nothing trivial, common, or mediocre about what has transpired on our home planet. Quite the opposite: the more we learn about other worlds, the more precious our world becomes. From what we have learned of life on Earth, the many steps needed from simple amino acids to self-aware multicellular creatures capable of pondering the meaning of existence, coupled with the "eerie silence" from extraterrestrial civilizations,[26] point forcibly toward our cosmic loneliness and not to a *Star Wars* universe populated with all sorts of smart creatures on distant worlds.

Of course, given the absence of evidence, we cannot conclude with confidence for or against the existence of alien life of any kind. Finding other life either directly through contact or indirectly through biosignatures on distant exoplanets is the only possible path forward toward clarity. To rephrase this point: not finding other life isn't proof of absence, only of rarity or of our inability to comprehend what this other life is. The Universe is vast, and our reach is limited. Still, we are the ones who know this. As we ponder the existence of life elsewhere, we enrich the Cosmos with our presence. The Universe has a history only because we are here to tell it.

CHAPTER SEVEN

LESSONS FROM A LIVING PLANET

I am the lover of uncontained and immortal beauty.
—Ralph Waldo Emerson, *Nature*

LIFE AND PLANET ARE ONE

The story of life on Earth is a story of Earth with life. Ever since life took hold here some 3.5 billion years ago, it interacted with the planet, transforming it as it was being transformed by the planet. Planet and life are like the ouroboros, the symbol of the serpent that bites its own tail, forming a closed circle. Once life takes hold of a planet, the two become an inseparable wholeness. We may think that humans are the first species to globally affect the planet, in our case much to the detriment of the biosphere. But that's incorrect. Current evidence suggests that photosynthesizing bacteria known as *cyanobacteria* were active about 2.2 billion years ago, dumping large amounts of oxygen into Earth's early atmosphere.[1] The oxygen filled the air and diffused through the oceans, creating an ozone layer as a byproduct. This layer in turn shielded Earth's surface from destructive solar ultraviolet radiation, protecting burgeoning life-forms and speeding up the process in a positive feedback loop. An oxygen-rich atmosphere changed the game of life. No other molecule allows for faster and more energy-releasing metabolic chemical reactions. As a consequence, organisms capable of using oxygen gained enormous evolutionary advantage. If an alien astronomer looked at Earth at this stage of its evolution, it would find a spectrum with a huge oxygen signature that essentially proclaims, "life is active here and photosynthesis rules." That's why we can't separate the story of a planet that harbors life from the story of life on that planet. Those ancient photosynthetic cyanobacteria changed Earth, giving rise to the Great Oxygenation Event. We and other forms of energy-demanding life are here largely because of this change.

The trouble is that oxygen is a highly poisonous gas when ingested in large quantities. (For example, divers and astronauts need to monitor their oxygen intake to avoid a condition called hyperoxia that causes severe tissue damage and death.) Cyanobacteria created an atmosphere supersaturated with oxygen, with nothing to make use of it. Any spark from lightning generated huge roaring fires that scorched Earth's surface. The situation stabilized only after organisms capable of breathing oxygen emerged. This is when the mitochondria made their triumphant entrance. They have their own DNA, indicating that they were once free-roaming organisms, microbes capable of processing oxygen. Then something remarkable happened. Primitive prokaryotic cells ingested mitochondria and somehow morphed into the larger and more sophisticated eukaryotic cells, the ones found today in the bodies of all animals and plants, other than eubacteria (such as cyanobacteria) and archaebacteria, a sort of intermediate group bridging the two types of cells.

In the expanse of 200 million years, about 1.9 billion years ago, eukaryotic organisms had evolved to the point of restoring the balance of the global carbon cycle, turning Earth into a viable platform for biodiversity to finally get going. This is when current estimates date the appearance of the last universal common ancestor (LUCA) of all eukaryotes, the most recent population of organisms from which all extant and extinct life-forms on Earth emerged, from primitive sponges and ferns to *T. rex*, fungi, and us. All life emerges from the same source.

Our collective Eve is a bacterium that lived some two billion years ago. The history of life on Earth shows that all life is connected, sharing the same seed in the distant past. We now understand that the details of life's evolution depend on the complex interplay between life and the planet. Randomness plays a big role, from the level of genetic mutations to global-scale cataclysms, such as massive volcanic eruptions and collisions with asteroids and comets. On top of those, there were severe glaciation events when the whole planet was covered in ice: snowball Earth. The very success of the cyanobacteria photosynthetic activity cleared the atmosphere of methane and carbon dioxide, creating a sort of inverse greenhouse effect: less greenhouse gases in the atmosphere led to rapid temperature decline, causing oceans to freeze from the poles to the equator. Ice covering the oceans locked huge swaths of cyanobacteria underneath, choking the production of oxygen. Life declined sharply, but persevered. Cyanobacteria survived near hot vents and hot water pools, like the ones we still see in Iceland and Antarctica. When the ice eventually melted, oxygen levels shot up, each snowball Earth event giving life a new series of creative boosts. Life's remarkable capacity to reinvent itself seems to thrive when it struggles the most for survival.

A dramatic snowball Earth event happened some 640 million years ago, setting the stage 530 million years ago for the Cambrian explosion, sometimes called the Big Bang of biology. A wild profusion of diverse animals, in the seas and on land, spread like fire in a geologically short period of time of about twenty million years, transforming the biosphere in astonishing ways. Complex multicellular animals emerged, forever changing the landscape of evolutionary diversity. Life's grand experiment played at the limits of the metabolically possible, spectacularly expanding its range into every conceivable niche the planet would offer.

Throughout this story, the fundamental rule for the success of emerging life-forms was and remained one and the same: adaptability to changing environmental conditions. If it gets too hot or too cold, if food supplies dwindle, if air quality is compromised, if a new predator becomes too efficient at hunting or prey at escaping, species will strive to adapt. Failure means extinction. Beneficial mutations may come to the rescue, but they are quite rare and their effects spread slowly, especially for more complex animals with longer gestation periods.

The story of life on any one world is unique. It will never be duplicated elsewhere, even on worlds that share very similar geophysical properties. Life on a hypothetical Earth 2.0, if it exists, will be very different from life here. Surely, certain evolutionary traits may reappear, such as the bilateral symmetry that we see in so many species on Earth. But life will be a new experiment every time, coevolving with the world that hosts it, unpredictable in its unfolding. No model starting with a bacterium on primal Earth could predict a lobster or a giraffe.

There is no plan behind what happens in the game of life. Our existence was never in the cards. If a single key episode in Earth's history hadn't happened, life would have taken a different path in an unknowable direction and we wouldn't be here. The most famous of these events is Earth's collision sixty-five million years ago with an eight-mile-wide asteroid in today's Yucatán peninsula in Mexico. That single event extinguished 75 percent of the then-living species, including the dinosaurs.[2] When this Manhattan-sized rock crashed into the ground at about forty-five thousand miles per hour, it punched a hole in our planet 110 miles wide and 12 miles deep, triggering a devastating combination of earthquakes, raging fires, and tsunamis, followed by a cloud of dust and debris that covered the Earth for months. As temperatures plummeted rapidly, plants depleted of sunlight died, causing a drastic disruption in the food chain. Species that could fly, burrow, or dive had huge survival advantages, including the small mammals that already existed then. After 60 million years of mutations and environmental changes, the first hominin species diverged from the apes, leading eventually to our species some 300,000 years ago. Compared with the age of the Earth, we have just arrived. Compressing 4.5 billion years into one day, we humans appear in this story only 5.7 seconds before midnight. We are the newcomers who think we own the place.

Before this cataclysmic event, the dinosaurs had existed for about 150 million years, the diversity of their many species slowly changing through mutations and selective pressures. They evolved for all these millennia and don't seem to have become intelligent enough to develop technologies or write poetry. By intelligence, I don't mean the crafty strategizing of a predator or the sophisticated underground burrows of prairie dogs and other animals, with tunnels connecting different clusters and chambers for babies and food storage. Instead, I mean a capacity for symbolic thinking and the ability to use fire to cook more digestible and nutritious food and other techniques to transform raw materials into tools that serve various purposes, from weapons to plowshares. The essential point here is that evolution is not a one-way street toward intelligence, even if intelligence is clearly an advantageous adaptive response to a changing environment. In other words, intelligence is not the inescapable result of a grand master plan for life. It emerged here by chance, and if it emerges elsewhere in the Universe, it will also be by chance.[3]

Life here spread through land, sea, and air, generating a staggering diversity of plants and animals over the past half a billion years; but intelligence flourished only recently, strongly suggesting that intelligent life must be rare in the Universe. How rare is difficult to quantify in any meaningful way, although, as we have seen, the complexity of each step leading from protocells to intelligent multicellular life, coupled with the deep silence of the Cosmos, points toward extreme rarity, if not to our uniqueness. This makes the existence of intelligence on this planet even more relevant, surprising, and remarkable. We should not take our existence here, or what it entails, for granted.

The story of life on Earth is a story of contingencies. The same will be true on any world where life emerges. There is no one path for life's evolution. There is no law of Nature that tells us that if life starts with bacteria, it will necessarily evolve to humanlike creatures. Intelligence is an evolutionary asset, of course; but this doesn't mean that life will *necessarily* evolve toward intelligent creatures, as 150 million years of dinosaurs and 2 billion years of single-celled organisms make amply clear. We shouldn't take the existence of *any* living creature lightly. Modern science tells us that there is nothing common or mediocre about *any* form of life. It tells us that no life-form is predictable. Evolution is like a map where the boundaries expand into untold possibilities. Copernicanism should never be applied to discussions about the possibility of life on other worlds. When it comes to life, inductive reasoning is the wrong tool.

THE UNIVERSE AWAKENS

The arguments presented so far in this book suggest that we must go beyond simple generalizations and extrapolations when discussing the existence of life, and especially the existence of intelligent life, in the Universe. The often-cited large number of exoplanets and their moons, the large number of galaxies, the validity of the physical and chemical laws across the known Universe, the number of Earthlike planets orbiting their host stars within their habitable zones—possibly in the billions in our galaxy alone—are simply not enough to say anything concrete about the existence of alien life. At most, these are necessary preconditions—far from sufficient—for life to flourish elsewhere, the first few steps of a very long staircase. Indeed, we long so much for alien life that we are perplexed by its absence, given that despite all these astrochemical properties that seem to favor life in the Cosmos, life turns out to be so elusive outside our world.

But we can and should turn the argument on its head. Instead of bemoaning the absence of life elsewhere and fearing the possibility of our cosmic loneliness, we should celebrate the life that blooms here and use this knowledge to retell our own story. Life on Earth is rare and precious, and we are the only species that knows this. Our planet is a living blue orb floating in the cold vastness of a Cosmos that cares nothing for our existence or for anything else. The Universe does not care. Through our cognitive capacities, the tenacity of our spirit, and our urge to know, we have been able to uncover many chapters of cosmic history, including parts of our own story on this world. This is our grand epic narrative of becoming, mythic in its range and significance, the story of the Cosmos and of life, and the realization that this story is our story. Life is the bridge between the eons, the shining beacon that has allowed us to look at the distant past to discover how our story and that of the whole Cosmos are deeply entwined. The Universe has a history only because we are here to tell it.

The history of life in the Universe can be told in terms of the steps needed for matter to have sequentially self-organized into structures of growing complexity, from elementary particles of matter to brains with a thick frontal cortex. I have organized this history into ages, which I call the Four Ages of Astrobiology.[4] The *Physical Age* starts with the Big Bang and the synthesis of the first atomic nuclei and goes all the way to the formation of stars and planets. The *Chemical Age* moves on to the synthesis of heavier chemical elements and simple biomolecules, leading eventually to the *Biological Age*, at least on Earth, where life emerged some

3.5 billion years ago and evolved into living creatures of growing complexity. Finally, about 300,000 years ago, during the *Cognitive Age*, our hominin ancestors branched out and became creatures with the capacity for symbolic thinking and complex languages. The history of life in the Universe begins with the origin of the Universe, unfolding for 13.8 billion years until life reaches the capacity for self-awareness with the Cognitive Age. When life is complex enough to tell its own story, the Universe awakens.

The Physical Age starts with the dawn of time, the Big Bang. We do not know how to make sense of the origin of all things—the problem of the First Cause. Our physical models are mechanistic, based on cause and effect, and cannot incorporate an uncaused cause without our making unprovable assumptions. The multiverse, as we have seen, is not a solution to the First Cause, and neither is any physical model that assumes this or that kind of geometry or this or that kind of matter existing at the very "beginning," including models that prescribe an ultrafast period of early expansion, known as inflationary cosmology.[5] The best that models can do is to make predictions that are compatible with the properties of the Universe we can measure. But compatibility is not a criterion for truth. Still, we *do* know how to tell the story starting a few fractions of a second after the "beginning," when space was filled with matter and radiation. The fact that we can do this is a spectacular achievement of modern science.

The Universe was born in time, and so was space. Time marked the cosmic expansion, the rate that the distance between any two fixed points in space increased. This cosmic expansion started 13.8 billion years ago, and it's still unfolding. We don't know the details of how matter, or what kinds of matter, emerged in this picture. For now, we have tentative models. But we do know that the expanding space was filled with matter furiously interacting with itself and with radiation, and that as space grew, matter and radiation cooled. ("Radiation" here means electromagnetic radiation, light and its invisible forms, like infrared and x-ray.)

The Physical Age tells the story of how this primordial soup evolved to become the protons and neutrons we find in the nuclei of all atoms, and how protons and neutrons combined to form the nuclei of the lightest atoms and their isotopes when the Universe was a few minutes old.[6] After that, the next big change happened when a different (and still unknown) kind of matter called dark matter started to clump together because of gravity, attracting normal matter to it as well. These clumps of dark and normal matter became, in the course of time, the first galaxies and stars. But before that, at about 380,000 years after the beginning, another great transition happened: the formation of the first atoms, when electrons and protons locked together to make hydrogen. From that time on, the Universe had hydrogen atoms, radiation, and a few floating nuclei, all being attracted to these growing clumps of dark matter. After tens of millions of years, gravity had squeezed these clumps so much that the hydrogen at their cores started to fuse into helium—a process known as *nuclear fusion*, the engine that powers stars. This is the age of the first stars, huge balls of hydrogen that lived short and dramatic lives. As they burned, they fused heavier and heavier chemical elements at their cores, before blowing off their innards in spectacularly powerful supernova explosions. Black holes were left behind, some becoming the seeds of nascent galaxies, all the while forming more stars that exploded and made more chemical elements. A billion years after the Big Bang, galaxies populated space, boasting stars of various sizes and gas clouds of mostly hydrogen that contained heavier chemical elements. With stars came planets, giving rise to the Chemical Age.

During the Chemical Age, stars made heavier chemical elements, and as they exploded, they spread their innards across interstellar space, seeding nascent stars and planets with the stuff of life. More complex chemistry emerged, including molecules of carbon dioxide, methane, and ammonia, some forged in young planets, some in gaseous regions called stellar nurseries, where stars are born. The Universe expanded and cooled further, galaxies aging and moving away from one another, with planets and some of their moons becoming progressively more enriched with heavier chemicals, including simple organics. Eventually, the stage was set for life to emerge somewhere in the Universe. About 9.3 billion years after the beginning, our solar system was born. Out of the fiery chaos of its origins, a rocky planet slowly took shape, third from the central star, a planet with lots of water and an early atmosphere rich in carbon dioxide. A billion years after, the first unicellular life stirred into existence, marking the transition from the Chemical to the Biological Age.

Whereas the Physical and Chemical Ages are still ongoing across the Universe as stars are born and evolve to their demise, the Biological and Cognitive Ages are known features of our solar system only, at least for now. The two life-related ages may have unfolded elsewhere (and may still do), but we will know this only if and when we discover extraterrestrial life.

The Biological Age may have started earlier elsewhere in the Universe, but so far we can tell its story only as it unfolded here.

Life took hold of this planet and persevered through the eons, despite tremendous upheavals and many near extinctions. In the game of life, challenges act as both destroyers and creators, given that they periodically reset environmental conditions that in turn redefine the needs for survival. As cataclysm followed cataclysm, some caused by outside influences, others by life itself, creatures mutated and changed, adapting as best they could. Survival is about efficiency and optimization, given the resources at hand. To remain alive every creature needs to strategize or, at the very least, be able to respond to environmental changes, including the presence of other creatures, friendly or hostile.

Life evolved in unpredictable ways, from simple one-celled microbes to microbial plants that used the Sun as their energy source while oxygenating Earth's atmosphere. In time, without a plan, life grew more complex, becoming multicellular, exploring all possible niches—water, land, and air, plant and animal—literally transforming its home planet into a living entity, a biosphere pulsating with creativity. More complex animals developed more complex survival strategies. About 6 million years ago, the first erect bipedal primates, the hominins, appeared in Africa. We now have evidence that by 3.3 million years ago some hominins, probably *Australopithecus*, fashioned primitive stone tools. The genus *Homo*—to which *Homo sapiens* belongs—dates back at least 2.8 million years. Our *Homo* ancestors grew progressively more adept in the usage of tools. *Homo erectus*, in particular, was an apex predator able to manipulate fire, an evolutionary game changer. They were hunter-gatherers who cared for their sick and may have developed art and even seafaring. Quite possibly, they communicated through a form of protolanguage.[7]

Being erect meant a shorter gestation period, which in turn meant babies that needed more care. What most animals had to do for hours, days, or months, the genus *Homo* had to do for years. Caring for babies, gathering food, and finding shelter called for group action, and sharing resources became essential for survival. This distinctive trait of our closest ancestors and our own species was deeply transformative. Groups banded together for long periods of time, probably for a lifetime, developing a sense of identity and belonging that ignited creativity and social rules for order. Rules had to be remembered and respected. Lasting bonds developed among members of the community. If we define cognition as the mental ability to acquire knowledge and understanding through thought, sensorial experience, and memory, some time after the emergence of the genus *Homo* life entered the Cognitive Age.[8]

Given the scarcity of information about the practices and living habits of Neanderthals and the earliest humans, it would be premature to try to date when this transition happened. One could, for example, attribute its beginning to the manipulation of fire. Taking a more pragmatic approach, one could likewise associate the Cognitive Age with the emergence of figurative art. Drawings of animals dating back as many as thirty-seven thousand years decorate the walls of the Chauvet Cave in France. Nicholas Conard, from the University of Tübingen in Germany (where Kepler went to school in the 1590s), found carved ivory figurines dating back forty thousand years. New radioactive decay methods date cave paintings in Borneo to forty thousand years ago. Current evidence thus indicates that from Indonesia to Germany, and at about the same time, early humans were representing their world through figurative art, probably to educate, entertain, and memorialize. Handprints on cave walls are a moving attempt to transcend time, to create something permanent or at least long-lasting. "We were here, don't forget us" their creators seem to be telling us, and we are grateful for their foresight. We will not forget them.

The dawn of the Cognitive Age opened endless possibilities. A species capable of symbolic thinking, of making up and telling stories, understands the passing of time and its own mortality. Once our ancestors began to tell their story, the Universe changed forever. As our storytelling grew in complexity and myths and philosophical musings joined our narratives, the Universe itself began to tell its story through our voices. Our stories became the cosmic story. Through our existence the Universe begot a mind.

Every culture, past and present, has its version of the origin of all things, of the world and of life. Creation myths tell how the land and the skies came to be and how animals and people emerged. They embody the values that define a culture. The Bible begins with the book of Genesis, establishing God as creator of all things. Ancient mantras from the Vedic period in India are considered the primordial rhythms of creation, preceding material forms. The Enuma Elish, the ancient Babylonian "Epic of Creation," starts with the line, "When above heaven had not [yet] been named."[9] Most of the time, these narratives describe the origin of everything through the action of a deity, or of many deities working together. This is how religions make sense of the problem of the First Cause. Only entities who are beyond space and time, beyond the constraints of the laws of Nature and thus *super*natural, can create that which is within Nature, bound by the rules of growth and decay. Some creation narratives, as with the Maori of New Zealand, tell of the origin of all things coming from an indescribable urge to burst into being, without divine intervention. Others, like the Jains of India, consider the Universe to be eternal and thus beyond the karmic cycle of reincarnation many Buddhist traditions teach.[10]

Despite their differences, all creation stories express the same universal human awe as we face the mystery of our existence in a reality that transcends understanding. As Einstein once wrote, "What I see in Nature is a magnificent structure that we comprehend only very imperfectly and that must fill a thinking person with a feeling of humility."[11] The story of who we are is also the story of how the Universe became what it is. Our fascination with origins turned us into the cosmic storytellers, the ones who find meaning deciphering the Universe's secrets.

PART IV
THE MINDFUL COSMOS

CHAPTER EIGHT

BIOCENTRISM

> The natural world is the larger sacred community to which we belong. To be alienated from this community is to become destitute in all that makes us human. To damage this community is to dimmish our own existence.
>
> —Thomas Berry, *The Dream of the Earth*

A Universe without life is a dead Universe. A Universe without minds has no memory. A Universe without memory has no history. The dawn of humanity marked the dawn of a mindful Universe, a Universe that after 13.8 billion years of quiet expansion found a voice to tell its story. Before life existed, the Universe was confined to physics and chemistry, stars forging chemical elements within their entrails and spreading them across space. There was no purpose to any of this, no grand plan of Creation. Through the unfolding of time, matter interacted with itself, as gravity sculpted galaxies and their stars. The emergence of life on Earth changed everything. Living matter doesn't simply undergo passive transformations. Life is "animated" matter, matter with purpose, the purpose of surviving.[1] Ecotheologian Thomas Berry wrote, "The term *animal* will forever indicate an *ensouled* being" (italics in the original).[2] Life is a blending of elements that manifests as purpose. This sense of purpose, this autonomous drive to survive, is what defines life at its most general. And in our world, the mountains, rivers, oceans, and air sustain every living being. Life elsewhere may be very different from life here. But if it exists, it must share the same urge to survive, to perpetuate itself in deep communion with its environment. The alternative, of course, is extinction. When life exists, it will struggle to remain existing. Life is matter with intentionality.

Life without higher levels of cognition doesn't know itself as living. It knows it needs to survive and will do what it can to remain alive, developing survival strategies with varying levels of complexity. It will search for food, eat when hungry, and sleep when tired; it will find or build shelter; it will protect itself and its young; it will fight to remain alive through force or strategy, as even plants are believed to do. Species evolved all sorts of remarkable tricks and weaponry to stay alive. Different animals have a range of emotions that may be quite expansive, although it is hard to truly understand what goes on in their psyches. Some may feel joy or sadness; some may help members of their species and even of other species, developing a true sense of companionship and caring. (Why else would we have pets?) But deep as their emotions might be, animals don't ponder the meaning of their existence. They don't have the urge to tell their stories and wonder about their origins. We do.

And what have we done with this remarkable ability? We became expert hunters and warriors, we became artists and storytellers, we worshipped gods and coveted love and power. We became a paradox, half beasts, half gods, capable of the most beautiful creations and the most atrocious crimes. We became the greatest lovers and the greatest murderers, thinking ourselves masters of this planet. We have turned our backs to the teachings of our ancestors and Indigenous cultures, who worshipped the land as their mother and the animals as their peers. We can tame much of what we fear, from fire to lions, and this power makes us giddy. But our ancestors knew, as we do, that we can't tame Nature. We can bend the course of rivers and raze forests, we can bring whole species into extinction, but we can't control the emergence of new diseases or stop cataclysmic events from killing us. We can kill wolves and tigers but not stop volcanoes from erupting. We are big and we are small, powerful and limited. Our success has lulled us into a false sense of confidence, leading us to believe that we are above Nature. But our planet, vast as it is, is limited, and it is responding to our voracity in ways that might destroy us or, at the very least, compromise the future of our species and countless others. We coevolve with Nature and can't extricate ourselves from its dynamics. To believe we can is our biggest mistake. Still, this is what we have attempted to do, creating a chasm separating us from the rest of Nature. We built huge cities and factories and country-sized mechanized agricultural monocultures, pushing wilderness to the unreachable fringes of the land. We consumed the planet's entrails, the oil and gas and coal, to feed our industrial growth. We lost touch with our evolutionary origins, with our roots in the wild, and we have forgotten who we are and where we came from. We have desecrated the land that sustains us, treating the world as our property.

This old narrative of the human has reached its end. The time has come for new humans, humans who understand that all forms of life are codependent, who have the humility to position themselves alongside all living creatures, and not above them. We have seen that this new narrative for humanity is founded on a confluence of cultures, merging Indigenous traditions with our growing scientific knowledge of the trillions of worlds around us. This new vision for humanity combines reason and spirituality, the material and the sacred, refusing to objectify the natural world. The fundamental tenet of this biocentric view is that a planet that holds life is sacred. And what is sacred must be revered and protected. A planet that holds life is profoundly different from the countless barren worlds spread across the vastness of space, marvelous as they might be. A planet that holds life is a living planet, and a living planet is where Cosmos and life embrace each other and create an irreducible wholeness. And of all the planets that may hold life in this galaxy and others, ours is a beacon of hope for being home to a species of storytellers.

The more we look to other worlds in search of signs of life, the more we realize how rare Earth is, how rare life is, how rare we are. We are the cosmic voice, capable of telling the cosmic story, and we need to rise above our destructive urges and our greed for immediate gratification to reorient our future. The story we have been telling until now, the Copernican narrative that we don't matter in the big scheme of things, that Earth is just a planet among trillions of others, is simply wrong. We matter because we are the only life-form that knows what it means to matter. We matter because we now understand how we are evolutionarily connected to every other life-form on this planet, descended as we are from the same bacterial ancestor. We matter because we know that life here is contingent on the whole cosmic history, from the properties of subatomic particles to the expansion of the Universe. We matter because we are how the Universe ponders its own existence. We matter because the Universe exists through our minds.

Moral rules are hardly universal. Those that to one group are terrorists, to another are freedom fighters. Values esteemed in one culture are criminalized in another. Different religions and political philosophies have different moral codes, and these differences have led to war and destruction across millennia. But the new understanding of how rare life is in this solar system and probably in most others should elevate one moral rule above all others. We no longer should think of the Universe only as a physical system. We must think of the Universe as home to life. The sacredness of a living planet is the central tenet of our post-Copernican narrative. We protect what is rare and precious. Life on Earth is rare and precious, planet and biosphere entangled in a single wholeness.

There is no life without Earth, but there is Earth without life. To transform Earth into one of our barren solar system neighbors would be the greatest crime humankind could commit against itself, against all life, against the Cosmos. Biocentrism is a vision for a morally aware humanity that celebrates and protects all forms of life as the only way to secure a healthy future for our project of civilization. It reaches beyond pre-Copernican human exceptionalism (we are the center of all Creation) and Copernican nihilism (we are nothing in the cosmic vastness), given that it weaves humankind into the web of life, the irreducible wholeness that enshrines the planet. Biocentrism presents humankind with a collective purpose, since, barring Earth experiencing a cataclysmic collision with a large asteroid, we alone have the power to preserve or destroy the biosphere. The alternative, inaction and neglect, will bring great suffering to all sectors of the population, especially—but certainly not exclusively—to those of weaker economic means and to our children and subsequent generations. The choice should be obvious.

This is our collective story, the story of a species that has learned to fashion raw materials into tools of exploration and objects of beauty, that has developed the ability to speak and tell stories about the experience of being alive, stories of love and loss, of war and heroic deeds, of triumphs and failures. The challenges we now face, products of our inability to build a sustainable relationship with the natural environment that supports us, we face together as a single species, as the human tribe. We fell into the hole we dug, but we can crawl out of it if we awaken to our cosmic role. If we truly matter, we should not erase our own legacy. We must reconnect to this planet and to all life on it with the humility and respect of the worshipper and not with the sword and rage of the slayer. This is the moral imperative of our age.

CHAPTER NINE

A MANIFESTO FOR HUMANITY'S FUTURE

> If you are a poet, you will see clearly that there is a cloud floating in this sheet of paper. Without a cloud, there can be no rain; without rain, the trees cannot grow; and without trees, we cannot make paper... So we can say that the cloud and the paper *inter-are*... Everything—time, space, the earth, rain, the minerals in the soil, the sunshine, the cloud, the river, the heat, and even consciousness— is in that sheet of paper. Everything coexists with it. To be is to inter-be. You cannot just *be* by yourself alone; you have to inter-be with every other thing... As thin as this sheet of paper is, it contains everything in the Universe.
>
> —Thich Nhat Hanh, *The Other Shore*

Ten thousand years of agrarian civilization have brought great prosperity and growth to a subset of humankind—those who own lands and the means of production—while also forging the path that brings death to our own and to the countless other living creatures that inhabit this planet.

This unstoppable prosperity and growth accelerated greatly after the advent of modern science and its technological offspring, the machinery of industrialization. The fuel for this growth came from the entrails of the planet—the oil, the gas, the coal—deeply transformed remains of life that thrived here millions of years ago. And also, regrettably and shamefully, from the carcasses of whales that have been decimated since the middle of the nineteenth century and continue to be so to this day.

This unchecked burning of fossil fuels, combined with population growth and an increase in life expectancy, and the consequent ever-expanding need for resources such as energy and food have taken a devastating toll on the natural environment. The material greed that feeds humankind's unquenchable thirst for more is suffocating the environment. Thus far, the formula has been simple: to keep on growing, we must keep on encroaching into the natural environment, believing that plants and animals are inferior life-forms without rights to live or to their own space, a space without the invasive human presence. Since the origins of the agrarian era, we have positioned ourselves as masters of the land, entitled to do with it as we please.

When the gods left the mountains, the rivers, and the forests and drifted upward to the heavens, Earth lost its enchantment, turning into an unprotected raw mass of rocks and trees and animals for us to use as we pleased, with the blessing of the gods above. We made ourselves owners of the natural world, believing we were creatures halfway between gods and animals. This fantasy is a dangerously mistaken inversion of hierarchies: Earth is primary, not the human. As Indigenous cultures have known for millennia, everything we do depends on the Earth and its resources, and everything we need comes from them. We no longer hunt and gather for our sustenance, but farming and mining still injure the Earth, as we cut it into manageable pieces, digging holes and razing mountains with impunity. This has been our way for more than ten thousand years—to treat the planet and its resources as our property, our right. Our current predicaments make it clear that this attitude is morally unjust and economically unsustainable.

We must reinvent ourselves, not by abandoning what we have accomplished, but by realigning our technological prowess and need for growth along a new moral stance, one that treats Earth and its biosphere as a sacred community to which we belong, not as masters but as equals to all other beings. Every voiceless creature has a right to live, just as we do. In our moral inconsistency, we shower our pets with abundant love but kill animals for food with disdain. We tend to our gardens as small temples but cut down forests with complete negligence. This is the behavior of a morally lost culture ruled by greed and not by compassion. Life alone knows life, and life feeds on life. Every living thing needs to eat; this reality cannot be changed. But *we* can transform it morally, by respecting and honoring what we kill, by mindfully replenishing what we harvest. We evolved to eat what we could find: meats, fruits, and roots. Our ancestors didn't have the luxury to choose what they ate. But now we have the know-how and means of production to greatly decrease our meat consumption and still have a full diet. We waste resources and food. We have replaced the efficiency of the hunter with the excesses of our productive machinery. We have made our planet sick, and a sick planet cannot support healthy lives.

At this stage, humankind is still an incoherent mass of dissenting tribes, most embracing value systems based on short-term thinking devoid of any deeper reflection of the mid- to long-term consequences of our choices. We march, many of us unwittingly, toward self-combustion. Many of us are unaware that our individual actions and choices play a role in our collective demise. Conveniently, we charge governments and corporations with being the agents of change. But history tells us that great transformations come from doers, from those who feel the need for change, those who have the courage to fight for the greater good. Think, for example, of great religious leaders like Jesus, Muhammad, or the Buddha and the countless martyrs who fell for their ideals. Think of Martin Luther King Jr., Mahatma Gandhi, and Nelson Mandela and their fight for racial freedom and equality. Or, from the birth of Western philosophy, think of Socrates, Plato, or the Stoics and their quest for living a life of meaning. And finally, think of the intellectual courage of people like Copernicus, Bruno, Galileo, and Kepler, who promoted their ideas about the Cosmos despite great adversity and danger.

Fortunately, the sacrifice here need not be a bloody revolution, but a readiness to make changes to the way we live, eat, and treat all humans, other life, and the planet. As revolutions go, this would be the most transformative to humanity, the first of its kind in our collective history, where we unite not as this or that tribe fighting against another, but as a whole species fighting for its survival and for the dignity of all living creatures. The banner of this movement is the sacredness of our planet. We fight against our past to create a new future.

Change is beginning to be felt, although it is still scattered and localized. A new wind of awakening is blowing, as climate change swirls into devastating storms and droughts, causing disease and famine across the globe, amplifying social injustices due to power differentials and economic disparity. With each day more people are growing in empathy, rejecting the inevitability of the suffering of other humans and of animals and plants.

There are two major obstacles to change, two great walls to be brought down. The first is our current narrative that places humankind above all other forms of life, as masters of the planet. The origins of this distorted view have roots in the monetization of the land, when someone decided that a piece of land has financial value attached to its ownership. Since then, we have believed that we own chunks of the planet that we call home. But even if we, like so many animals, need shelter to survive and thrive, the first home we should think of is our planet, the home we share with all other creatures. Earth is our primary home, the home that makes life possible. Take away our oxygen-rich atmosphere, take away our protective magnetic field, take away the slow drift of tectonic planets, take away our large Moon, take away our shielding ozone layer, and life as we know it wouldn't exist. *Everything* that we do follows from the premise that Earth gives us the possibility to exist, including owning our homes. The point here is not to abolish land ownership but to understand that the piece of real estate we happen to own doesn't *belong* to us; we are borrowing it temporarily from the planet. The land we now live on will be here long after we and our descendants are gone. The concept of land ownership is a fabrication of the economic system we have adopted, an ephemeral fantasy based on human hubris. To recalibrate our current values, we must acknowledge these facts and reverse our beliefs by placing Earth above the human, being grateful for our existence in this bountiful world.

Modern astronomy, as we have seen, can help us reframe this narrative, as we now understand the place of Earth in the Cosmos as the rare home of a species that is mindful of its existence, capable of telling its own story and of linking it to the story of the Universe. Our post-Copernican, biocentric narrative places life as the supreme cosmic achievement and humankind as its voice and memory. The more we learn about the Universe, the more we understand why we matter. We are messengers from the stars and for the stars.

The second obstacle is our fixation on the material at the expense of the spiritual. We need to awaken to an age of secular spirituality, one that embraces various forms of belief and non-belief. Secular spirituality is nondenominational, unrelated to supernatural notions of immortal souls and spirits. Humans are spiritual beings. We long for meaning and transcendence, for moments and experiences that transport us to emotional states that are encounters with the sacred. Some connect with the sublime through traditional prayer, some explore mountains and deserts, some play music and dance, some practice meditation, some write poetry or paint, some practice martial arts, some use psychedelic journeys to expand their minds and hearts, some merely sit in the quiet of their rooms. The choices are subjective and culturally driven. But the longing for awe as a path for self-growth and transformation is universal, a shared need to connect with the mystery of who we are.

The Enlightenment opened countless doors to what we can do through the diligent use of our rational faculties. We have prospered greatly as a result. The fuel for this unprecedented growth in human history came from Earth and its resources. How many of us stop to think about this obvious fact? As the brilliant philosopher and theologian Simone Weil made clear, the mechanization of the economy made sure to erase the human soul.[1] No one cutting down thousands of acres of pristine forest, boring holes in the ground for mining, or killing hundreds of food animals a day could be spiritually connected to Earth or to life. Most certainly not the ones ordering such actions or investing in the companies that implement them, even if sometimes unwittingly. At this juncture in human history, we must take responsibility for how our business investments reflect our values, or lack of them. Such devastation could be perpetrated only by those who consider the planet and life on it as useless things, without any value or rights. The severance of our spiritual bond to the planet made our unchecked growth possible. Nature changed from being a sacred realm to disposable garbage. This situation is now untenable, since this growth is choking our future.

With these arguments in mind, what measures can we implement to change the current course of civilization?

1. The core value of *biocentrism* is that a planet that hosts life is sacred, deserving of respect and veneration. We are part of the life collective, codependent and coevolving with the whole of the biosphere.
2. The conceptual basis for this change of mindset comes from the scientific realization that *life is a rare event in the Universe and that Earth is a rare planet*. There may be life elsewhere, even intelligent life. However, for all

practical purposes, given the vast interstellar distances and the lack of evidence supporting the existence of extraterrestrial life, we are alone and must rise to the task of rescuing our project of civilization from its current disastrous path.

3. This is a revolution geared toward the *spiritual reawakening of humanity*, a denomination-free spirituality centered on the reconnection of each of us to the land and to all forms of life.

4. There is nothing naive or innocent about *a movement that combines science and secular spirituality*. What is naive and innocent is to continue believing that things can remain as they are and all will be well, or that there is nothing we can do, or that science alone will save us. Science is surely an essential tool for our collective future, but without a passionate connection to Earth and life as a driver for change, without a sense of belonging to the world, without a firm belief that we can collectively change as a species, science will continue to be mostly used to expand our control over the natural environment without any moral concern, as it has been since the Industrial Revolution and before. For science to be a force for our collective good it must align with biocentric values that reflect our spiritual reconnection with Earth and the biosphere. This is beginning to happen, but not fast enough.

5. *Each individual has a role to play*. This role involves sacrifices that are entirely distinct from those of a bloody revolution. Instead of paying with our lives,

we celebrate and preserve life and align our values and actions according to three principles: the LESS approach to sustainability, the MORE approach to engagement with the natural world, and the MINDFUL approach to consumerism.

» *The LESS approach to sustainability:* Individuals should critically examine what they eat, how they use energy and water, how much garbage they produce, and how they dispose of it. The approach should focus on LESS: less meat, less energy, less water, less garbage.

» *The MORE approach to engagement with the natural world:* Whenever possible, individuals should engage more with Nature. If forests and natural parks, oceans, mountains, and trails are not available, then take walks on waterfronts, explore city squares and parks, and plant gardens at home. Schools and families can take children on hikes and excursions into the woods and organize visits to positive examples of environmentally mindful farming and industrial practices. Such initiatives will greatly help change the overall mindset that Nature is expendable. The approach should focus on MORE: more awareness of the life around us, big and small; more gratitude for the planet that allows us to be alive and flourish; more kindness to all forms of life.

» *The MINDFUL approach to consumerism:* Consumers have the power to shape corporations and their policies. The logic is simple: if consumers don't buy, corporations don't sell and are forced to change their practices. United, consumers have great power to make changes. Individuals should be mindful of the companies they buy products from. Do these companies align with biocentric values? Do they strive to minimize their carbon footprint? Do they promote best practices with high ethical environmental values? Do they promote inclusivity and gender equality in their hiring practices? Do they have a philanthropic practice, giving back to society and the planet? Are they mindful of their production chain, and do they respect their workers? Do they create a partnership with their clients, or do they see clients as targets for their marketing "campaigns"? As more people buy products from companies that embrace a sustainable and forward-thinking environmental ethic, the prices of those products will go down and become affordable for more consumers. **Consumers of the world unite so Nature can win!**

6. Independently of political or religious affiliation, *all schools should add the history of the Cosmos and of life on Earth to their curricula at all levels,* deepening instruction details accordingly for older students. This science-and-humanities-based cosmic narrative should never be part of a liberal versus conservative rhetoric. A change in mindset begins with a change

in moral orientation. If humankind is to change its relation to the planet and to life, this change needs to be nourished in all classrooms and dining rooms, promoted by teachers and families. The reshaping of our collective future starts with learning the story of life's past, its unity, and our deep link to the history of the Universe.

Natural spirituality is a practice, a pursuit that needs the engagement of body and mind with the land, a tactile, visceral bonding of the whole of us with the whole of the planet. Once each of us internalizes this bond, and experiences the inter-being that embraces all that is, our outlook changes. We begin to regret our past ways and ready ourselves to enact the changes to come. We have the opportunity and the privilege to treat our Earth with gratitude and awe. When we turn our attention to the wonders of our world we feel the inviting embrace of the life collective.

EPILOGUE

THE RESACRALIZATION OF NATURE

There is a sadness in concrete that makes the soul sink. The drab dullness of gray walls towering over our heads, the grime of sidewalks studded with gum and spit—symbols of a human presence that does not care for the ground it steps on or the skies above. The stale smells and vapors emanating from the underground, a stink of hot steel and sweat. The quiet traffic of downturned faces crisscrossing the roads, busy and crestfallen. There is the occasional smile, the salute from a stranger, as if to remind us that underneath all this there is still a human filled with dreams and aspirations. And there is also an austere beauty in the city, sometimes, when the architect is intent on creating an object to behold. That could be the feeling of someone walking down an avenue in New York, shielded from the sky and Sun by lit-up skyscrapers. Or admiring Chicago's skyline from the lakefront, or Paris in the spring. But it's in the parks that people congregate, it is the river that cuts and nourishes the city, it is the light reflecting on fountains and ponds that adds vitality to the sights. We won't abandon the cities, of course. But we can strive to rebalance our lives, embracing the natural environment in creative ways, opening cities to the green and the blue we have pushed away.

Our minds tend to create a fantasy of the exact as we build cities, a comfort of the precise, of straight lines and sharp angles, of perfect curves and cones, the likes of which we never find in the wilderness. There, far from the fabricated geometry of the city, lines are jagged and uncertain, rocks and leaves are asymmetrical, straying from perfection sometimes just enough to enhance the beauty of the unexpected. The softness of natural forms invites us into the landscape, to the welcoming embrace of a forest or mountain. We evolved in the wilderness but have since turned it into a stranger, often a terrifying and distant one. We have strayed so far from our origins that we perceive the untouched landscape as menacing and foreign. Humans built cities to push Nature out, and now we feel at home within concrete walls, straight and predictable. Nothing in an apartment or house moves by itself. We close the shades of our windows to block the sunlight, preferring instead to illuminate our interiors with artificial lights that imitate the Sun. The deeper we wall ourselves in, the more detached we feel from the natural environment, and the more objectified it becomes.

Yet every life-form carries the mountains, rivers, oceans, and air within, a moving, breathing expression of the world manifest as being. This is the community of the living with the nonliving, being and matter as one, the sacred bond of existing. To sever this bond is to decree our own oblivion. We cannot survive believing we are above Nature.

A world created by a god is sacred. A world described exclusively by science, a world of chance and causation, is not. Wounding a god-created world is sacrilege. Wounding a world of chance and causation is just part of survival. When science banned the gods from the world, it opened it to human plunder. We see this clearly as we compare the values of Indigenous cultures—their respect for the land and for everything and everyone they share it with, living and not—with those of secular industrial society, with no spiritual attachment to the world. We don't want to bring the gods back to Earth, but we do need to reestablish the spiritual connection between humans and Nature. This is where Indigenous wisdom connects with modern scientific knowledge to illuminate the path ahead. We need to resacralize the world so we grow to respect it with renewed passion. "Sacred" doesn't mean a realm haunted by supernatural divine presences. It means a realm that enables us to engage with the mystery of existence, to be awestruck as we connect with the sublime, to worship the world as a temple, to humbly bow to natural powers vastly beyond our control.

The full realization of our humanity will blossom when, together as a species, we embrace the life collective as one. This is the moral imperative of our era. This is our sacred mission.

ACKNOWLEDGMENTS

We often don't realize that the people who are the most inspiring, most transforming in our lives are the ones closest to us. Over the years, as I gathered the ideas that became this book, I kept going back to my childhood, to some of my early fragmentary memories of my parents and brothers, sharing moments of bliss in the gardens my father cared for with such love and tenderness. He poked the dirt with magic fingers, bringing forth life from tiny seeds, invisible threads of nothing that exploded into thousands of flowers and fruits in spectacular tropical profusion. "A seed contains the tree that it becomes and the fruits that it gives us," he would say. A court of insects and birds paid daily homage to this bounty, as I, bewildered, tried to learn the secrets of what life entails, a connection beyond words, a bridge between hearts, a sharing of visions. Even in the saddest of times, possibly to counter them, my father would go out to his gardens to coax life into the world. I wouldn't be me without you.

My brothers and their families have been a steady antidote to the hardships that befall us at some points in our lives, and I couldn't be more grateful for your love, for your laughter, for your teasing, and for the endless debates about everything and nothing—all argued with passionate conviction.

My children continue to inspire me as they grow, each a wonderful, creative, and kind-hearted human, each moving along their path in life with a deep sense of moral justice and passion for knowledge. I am truly blessed to have you in my life.

My wife, Kari, brightens every day of my life with her vibrant, noble, and profoundly generous self, a companion one is allowed to have only in dreams. I feel as if I have fooled the gods to have you by my side.

My dear friends Mauro, Everard, Adam—you help me make sense of the world, when nothing seems to, in ways you don't even know. I only regret we don't find more time in our lives to spend together.

My agent and friend, Michael Carlisle—for believing in this project from the start, and for pushing hard for it to become a reality.

My heartfelt thanks to Bill Egginton, for reading the manuscript and for his inspiring feedback and wisdom; to Jeremy DeSilva, for his knowledge about our distant ancestry; and to Mary Evelyn Tucker, for introducing me to the wonderful world of Thomas Berry.

My editor, Gabriella Page-Fort—it's hard to believe you came late to this project. For clearly you were meant to be an integral part of it from the very beginning. Mysterious and unknowable connections happened to make our paths align, and I couldn't be more grateful for them and for the wisdom you brought to this project. And to my copyeditor, Jessie Dolch, for her wise help and suggestions.

NOTES

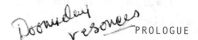

PROLOGUE

1. Here is a very incomplete list: Rachel Carlson, *Silent Spring*, 40th anniversary ed. (New York: Houghton Mifflin, 2002); Elizabeth Kolbert, *The Sixth Extinction: An Unnatural History* (New York: Picador, 2015); Toby Ord, *The Precipice: Existential Risk and the Future of Humanity* (New York: Hachette, 2020); James Lovelock, *The Revenge of Gaia: Earth's Climate Crisis and the Fate of Humanity* (New York: Basic Books, 2007); Bill McKibben, *Falter: Has the Human Game Begun to Play Itself Out?* (New York: Henry Holt, 2019); Bill McKibben, *The End of Nature* (New York: Random House, 1989); Benjamin von Brackel, *Nowhere Left to Go: How Climate Change Is Driving Species to the Ends of the Earth* (New York: The Experiment, 2022).

CHAPTER ONE
COPERNICUS IS DEAD! LONG LIVE COPERNICANISM!

1. All quotations from Copernicus are from *On the Revolutions of the Heavenly Spheres*, preface by Andreas Osiander (London: Macmillan, 1978).

CHAPTER TWO
DREAMING UP THE COSMOS

1. G. S. Kirk and J. E. Raven, *The Presocratic Philosophers: A Critical History with Selected Texts*, 1st ed. (Cambridge: Cambridge Univ. Press, 1957),

henceforth K&R in the text, followed by page number. There is a second edition of this work, which includes M. Schofield as an author, but I have the original edition, which, since my teenage years, I have carried with me wherever life has taken me.

2. For a technical work, see Stephon Alexander, Sam Cormack, and Marcelo Gleiser, "A Cyclic Universe Approach to Fine Tuning," *Physics Letters B* 757 (2016): 247–250.

3. Mary-Jane Rubenstein, *Worlds Without End: The Many Lives of the Multiverse* (New York: Columbia Univ. Press, 2014), 42.

4. Epicurus quoted in Rubenstein, *Worlds Without End*, 251.

5. Lucretius, *The Nature of Things*, Penguin Classics, trans. A. E. Stallings (London: Penguin Books, 2007), 5:218–222.

6. Stephen Greenblatt, *The Swerve: How the World Became Modern* (New York: W. W. Norton, 2012).

7. Lucretius, *Nature of Things,* 5:238–240, 243–245.

8. Rubenstein, *Worlds Without End*, 54.

9. John Muir, *My First Summer in the Sierra* (New York: Penguin Books, 1987), 157.

10. Isaac Newton, *The Principia: Mathematical Principles of Natural Philosophy*, trans. I. Bernard Cohen and Anne Whitman (Berkeley: Univ. of California Press, 1999), 943, 940.

11. Marcelo Gleiser, *A Tear at the Edge of Creation: A Radical New Vision for Life in an Imperfect Universe* (New York: Free Press, 2010).

12. Marcelo Gleiser, *The Island of Knowledge: The Limits of Science and the Search for Meaning* (New York: Basic Books, 2014).

13. A note for the experts: The justly celebrated unification of the weak and electromagnetic forces known as the electroweak theory is not a true unification. A true unification of two forces must have a single coupling constant related to a single gauge symmetry group. The electroweak theory retains two coupling constants and two symmetry groups. The

unification here refers to the fact that the gauge bosons of the weak force, which are massive at low energies, behave effectively as being massless at higher energies, as does the photon from electromagnetism (at all energies). So, at higher energies the theory has four effective massless gauge bosons. This is conceptually different from a grand unified theory, where three forces (electromagnetism, weak and strong nuclear forces) would presumably be unified under a single gauge group. Unfortunately, since being proposed in 1974, no signs of such grand unification have been observed.

14. Adam Frank, Marcelo Gleiser, and Evan Thompson, *The Blind Spot: Why Science Cannot Ignore Human Experience* (Cambridge, MA: MIT Press, forthcoming).

15. The real date of publication of *Micromégas* is unknown, but it is estimated to be 1752, given by the Kehl edition. A recent edition is Voltaire, *Micromégas and Other Short Fictions* (Penguin Books, London, 2002). The short story is available online at Project Gutenberg, https://www.gutenberg.org/files/30123/30123-h/30123-h.htm.

16. Marcelo Gleiser, "Pseudostable Bubbles," *Physical Review D* 49 (1994): 2978. This is the paper where I first proposed the name oscillons.

17. Scientific induction is powerful but fallible. A famous example is the swan. Up to the seventeenth century, any European would have agreed that all swans are white, as all swans observed up to then were white, so by induction, people generalized this to all swans. This was the accepted "truth" until the Dutch explorer Willem de Vlamingh found black swans in Australia in 1697.

18. Werner Heisenberg, *Physics and Philosophy: The Revolution in Modern Science* (New York: Penguin, 2000), 25.

19. Jorge Luis Borges, "On Exactitude in Science," in *Collected Fictions*, trans. Andrew Hurley (New York: Penguin, 1999), 325.

20. Anthony Aguirre and Matthew C. Johnson, "A Status Report on the Observability of Cosmic Bubble Collisions," *Reports of Progress in Physics* 74 (2011): 074901, https://arxiv.org/abs/0908.4105; Matthew Kleban,

"Cosmic Bubble Collisions," *Classical and Quantum Gravity* 28 (2011): 204008, https://arxiv.org/abs/1107.2593.

21. "These matters being very extraordinary, will require a very extraordinary proof," Benjamin Bayly, *An Essay on Inspiration: In Two Parts* (1708; reprint Whitefish, MT: Kessinger, 2010). Bayly's sermons concerned the veracity of miracles as a way for God to reveal his presence to rational men in ways that were irrefutably true. For this, the miracles had to be beyond the possible or accidental.

22. For the experts, a current model where matter and energy expand and contract while space mostly expands with very small contraction evokes a quintessence-like scalar field with changing behavior and a negative exponential potential and an "appropriate modification of Einstein gravity at high energy densities near the bounce or stress-energy that violates the null energy condition or both." See Anna Ijjas and Paul J. Steinhardt, "A New Kind of Cyclic Universe," *Physics Letters B* 795 (2019): 666–672, https://arxiv.org/pdf/1904.08022.pdf.

23. This issue is raised in a recent paper of mine submitted in collaboration with Sara Vannah and Ian Stiehl, "An Informational Approach to Exoplanet Characterization," *International Journal of Astrobiology*, preprint, submitted June 27, 2022, https://arxiv.org/abs/2206.13344.

CHAPTER THREE
THE DESACRALIZATION OF NATURE

1. Ailton Krenak, *Ideas to Postpone the End of the World* (Ontario: House of Anansi Press, 2020), 3.

2. Of course, many Indigenous cultures also settle into agrarian societies and trade goods with each other. However, their relationship with the land is one of sustainability and respect, as opposed to one of exploitation and neglect.

3. According to recent scholarship—albeit somewhat controversial, see, e.g., David Graeber and David Wengrow, *The Dawn of Everything: A New History for Humanity* (New York: Macmillan, 2021)—even

hunter-gatherer communities experimented with several kinds of social order and hierarchical control, some of these later incorporated into larger agrarian societies. For us, what matters is the shift from a spiritual to an exploitative stance for how humans related to the land.

4. Thomas Berry, *Evening Thoughts: Reflecting on Earth as a Sacred Community*, ed. Mary Evelyn Tucker (San Francisco: Sierra Club, 2006).

5. Why else would God reveal himself to Moses as a burning bush—a clear statement of the impossible in the earthly realm?

6. In fact, Copernicus dedicated his book *On the Revolutions* to Pope Paul III. His fiercest critics were Lutherans, such as Andreas Osiander (see chapter 1) and Martin Luther himself, who once referred to Copernicus as "a new astrologer who wants to prove that the Earth moves and goes around instead of the sky . . . The fool wants to turn the whole art of astronomy upside-down." Quoted in Noel Swerdlow and Otto Neugebauer, *Mathematical Astronomy in Copernicus "De Revolutionibus,"* 2 vols. (New York: Springer, 1984), vol. 1, 3.

7. Details would depend on the objects' initial position (and direction) and initial speeds. For example, a gun shooting a cannonball could be aimed at different heights and the cannonball propelled forward with different speeds, resulting in different (parabolic) paths. But the force is always Newton's universal gravity between two masses M_1 (the cannonball) and M_2 (the Earth). Or, for a planet, between the planet and its host star.

8. Isaac Newton, *The Principia: Mathematical Principles of Natural Philosophy*, trans. I. Bernard Cohen and Anne Whitman (Berkeley: Univ. of California Press, 1999), 943.

9. Isaac Newton, "Four Letters to Richard Bentley," letter from February 25, 1692, in I. Bernard Cohen and Richard S. Westfall, eds., *Newton* (New York: W. W. Norton, 1995), 336–337.

10. Newton, *The Principia*, 942.

11. Isaac Newton, "Four Letters to Richard Bentley," letter from December 10, 1692, in Cohen and Westfall, *Newton*, 332.

12. John Maynard Keynes, from Address to the Royal Society Club (1942), quoted in Alan L. MacKay, *A Dictionary of Scientific Quotations* (London: Institute of Physics Publishing, 1991), 140.

CHAPTER FOUR
THE SEARCH FOR OTHER WORLDS

1. William Wordsworth, "Lines Composed a Few Miles Above Tintern Abbey, on Revisiting the Banks of the Wye During a Tour. July 13, 1798."

2. Robert Macfarlane, *Mountains of the Mind: Adventures in Reaching the Summit* (New York: Vintage, 2004), 157.

3. Isaac Newton, *The Principia: Mathematical Principles of Natural Philosophy*, trans. I. Bernard Cohen and Anne Whitman (Berkeley: Univ. of California Press, 1999), 938.

4. Wordsworth lived in Somerset until 1798, when he moved (with Coleridge) to the Lake District.

5. Earlier sightings of Uranus date back to at least 128 BCE, when the Greek astronomer Hipparchus recorded it as a star in his star catalogue.

6. J. L. E. Dreyer, *The Scientific Papers of Sir William Herschel* (London: Royal Society and Royal Astronomical Society, 1912), 1:100.

7. Quoted in Edward S. Holden, *Sir William Herschel: His Life and Works* (New York: Charles Scribner's Sons, 1880): 85.

8. Reflecting telescopes capture light with one or more curved mirrors that is then focused and directed to an eyepiece or another instrument. Isaac Newton invented them to improve on the distortions (chromatic aberration) caused by the then-common refracting telescopes such as Galileo's, which was made of two lenses.

9. I offer an in-depth exploration of this point in my book *The Island of Knowledge: The Limits of Science and the Search for Meaning* (New York: Basic Books, 2014).

10. William Herschel, "On the Power of Penetrating into Space by Telescopes; with a comparative determination of the extent of that power in natural vision, and in telescopes of various sizes and constructions; illustrated by select observations," *Philosophical Transactions of the Royal Society* 90 (Dec. 1800): 49–85.

11. George Basalla, *Civilized Life in the Universe: Scientists on Intelligent Extraterrestrials* (New York: Oxford Univ. Press, 2006).

12. To get to the Moon, Kepler's traveler used magic, which he feared could further implicate his mother, who had been accused of witchcraft and was nearly burned at the stake. She narrowly escaped this horrendous fate only because of her son's clever legal defense.

13. Christiaan Huygens, *Cosmotheoros: The Celestial Worlds Discovered, or, Conjectures Concerning the Inhabitants, Plants and Productions of the Worlds in the Planets* (London: Timothy Childe, 1698), available at https://webspace.science.uu.nl/~gent0113/huygens/huygens_ct_en.htm.

14. Bernard Le Bovier de Fontenelle, *Conversations on the Plurality of Worlds*, trans. H. A. Hargreaves (Los Angeles: Univ. of California Press, 1990), 45

15. Fontenelle, *Conversations*, 72.

16. Fontenelle, *Conversations*, 11.

17. From the French, "Le planète, dont vous avez signalé la position, réellement existe." Davor Krajnović, "The Contrivance of Neptune," available at https://arxiv.org/ftp/arxiv/papers/1610/1610.06424.pdf. This article also explores the huge controversy surrounding Neptune's discovery, also claimed by the British astronomical community at the time. Current consensus attributes it to Le Verrier and Galle.

18. The forthcoming book *The Blind Spot: Why Science Cannot Ignore Human Experience* (Cambridge, MA: MIT Press), written with my colleagues Adam Frank and Evan Thompson, examines in great detail the importance of experience in the scientific enterprise.

19. Eugene Wigner, "The Unreasonable Effectiveness of Mathematics in the Natural Sciences," *Communications in Pure and Applied Mathematics* 13, no. 1 (Feb. 1960): 1–14.

20. Very slowly indeed. It moves at the rate of 5557 arc seconds *per century* (equivalent to 1.54 degrees), of which 43 arc seconds are due to the effects from the Sun (as explained in Einstein's general theory of relativity), and 5514 are due to the gravitational tug from other planets. Newtonian physics could account for only the 5514 arc seconds from normal gravitational tugging. The extra 43 arc seconds were what Le Verrier wanted to attribute to planet Vulcan and what Einstein explained with his new theory of gravity. Recall that 1 arc second is equal to $1/3{,}600$ of a degree—a tiny angle.

21. Thomas Levenson, *The Hunt for Vulcan: . . . And How Albert Einstein Destroyed a Planet, Discovered Relativity, and Deciphered the Universe* (New York: Random House, 2016).

22. Since light from distant stars was visible from Earth, the ether had to be perfectly transparent. It couldn't offer any friction or it would interfere with planetary orbits. It had to be quite rigid to sustain the propagation of light waves at 186,000 miles per second. A truly magical medium that, alas, doesn't exist.

23. H. G. Wells, *The War of the Worlds* (New York: Tor, 1988), 187.

24. By the end of its mission, *Perseverance* will have collected forty-three samples of Martian rocks and soil to be shipped back to Earth. Involving multiple launches, multiple spacecrafts, and dozens of government agencies, the Mars Sample Return program is as ambitious as it is spectacular—with plans to return the samples to Earth by the early to mid-2030s for analysis.

25. The book *Chasing New Horizons: Inside the Epic First Mission to Pluto* (New York: Picador, 2018), by mission leader Alan Stern and astrobiologist David Grinspoon, is a must-read.

26. The word "world" tends to be used quite freely in astronomy and in popular culture. I'm using the word "worlds" to include all celestial

objects massive enough to have a gravitational pull that holds small creatures to the surface. In practice, this means worlds are objects with a large enough escape velocity, that is, the velocity it would take to escape their gravity and fly out into space. For the experts, the escape velocity in kilometers per hour is $V_e = 4.2 \times 10^{-5} (M/R)^{1/2}$ km/h, where M is the mass in units of kilograms and R is the average radius in units of meters. As an example, the dwarf planet Ceres has a mass 1.3 percent of the Moon's mass and a radius of 469.73 kilometers, giving its escape velocity as 1,890 kilometers per hour, about 1.5 times the speed of sound.

27. It's not really belt shaped, since space has three dimensions and not two. The habitable zone is more like a thick shell, even if we refer to it as being belt shaped.

28. For comparison, Titan is larger than Mercury and our Moon.

29. The moon Tethys is tinged slightly blue from infalling ring materials, while the trojan moons Telesto, Calypso, Helene, and Polydeuces have smoothed surfaces from materials that accumulate as they cruise along the ring plane in their orbits.

30. Here is an unfairly biased list of my favorites: Adam Frank, *Light of the Stars: Aliens Worlds and the Fate of Earth* (New York: W. W. Norton, 2019); David Grinspoon, *Lonely Planets: The Natural Philosophy of Alien Life* (New York: Ecco, 2004); Paul Davies, *The Eerie Silence: Renewing Our Search for Alien Intelligence* (New York: Houghton Mifflin Harcourt, 2010); John Gribbin, *Alone in the Universe: Why Our Planet Is Unique* (New York: Wiley, 2011); and Caleb Scharf, *The Copernicus Complex: Our Cosmic Significance in a Universe of Planets and Probabilities* (New York: Farrar, Straus and Giroux, 2014).

31. Objects with masses smaller than that are substellar, that is, not massive enough for nuclear fusion. With masses between thirteen and eighty times that of Jupiter, they are called brown dwarfs. They are too light to sustain the fusion processes at their core to ignite as a star. They are, in a sense, failed stars.

32. Just to confuse people, physicists use "blue" for hot and "red" for cold, the opposite of how we indicate temperature for, for instance, tap water. This choice is related to the colors of the rainbow, which is made of light with different wavelengths. (To visualize a wavelength, picture yourself throwing a rock in a pond. You will see concentric waves moving outward from the point of impact. The distance between the waves is the wavelength.) Light waves at the blue end of the spectrum have shorter wavelengths and carry more energy than those at the red end. This energy can be associated with temperature; hence, the connection. This also includes electromagnetic waves, which are invisible to the human eye, like infrared or ultraviolet radiation. Even though the energy associated with an electromagnetic wave is proportional to the square of the electric and magnetic fields in a given volume of space, when describing the interaction of light with atoms and subatomic particles, physicists also refer to the energy of the associated *photons*, the particles identified with the smallest "light particles" corresponding to the wave's wavelength. Here is the formula for the energy of a photon: $E = h\,c/L$, where E is energy, h is Planck's constant, c is the speed of light, and L is the light's wavelength. Since h and c are constants of Nature (they don't change), the smaller the wavelength L, the more energy the photon carries.

33. Readers who would like to go deeper into the science but have no technical background are encouraged to check out one of many introductory textbooks on the subject. One of my favorites is *Life in the Universe* by Jeffrey Bennett and Seth Shostak (Boston: Pearson, 2017).

34. Christopher P. McKay, "Requirements and Limits for Life in the Context of Exoplanets," *Proceedings of the National Academy of Sciences (PNAS)* 111, no. 35 (2014): 12,628–12,633.

35. There is still no consensus among the scientific community as to when life took hold on Earth. Estimates vary between 4.0 and 3.5 billion years ago. The difficulty is that as we go farther into the past, fossil records become harder, or impossible, to find. Determining whether a 4.0-billion-year-old rock carries hints of life depends on very complex

interpretations of the chemical compounds found in the rock that may or may not be derivative of primitive metabolic activity on primal Earth. We do know that life was present by 3.5 billion years ago, about 1 billion years after Earth formed. In any case, on a planet like ours, we can say that life takes at least a few hundred million years to take hold. This gives us a rough estimate of what kinds of stars could be good candidates for hosting life. Types O and B stars are essentially ruled out, and type A stars are borderline interesting.

36. This is not to say that there aren't images already. We can see a few exoplanets forming around nascent stars, using, for example, the Very Large Telescope from the European Southern Observatory. But the level of detail is far from what is needed for a more careful analysis of any planet's properties. (The interested reader can find images taken by the Hubble Space Telescope online at various websites.)

37. Why "blue" and "red" shifts? The visible light spectrum spans the colors of the rainbow, from violet (high-frequency waves) to red (low-frequency waves). So, a light source moving toward an observer will appear bluer, whereas one receding from the observer will appear redder. Of course, this remains true for the invisible parts of the electromagnetic wave spectrum, from the very-low-frequency radio waves to the very-high-frequency gamma rays. This way, astronomers can measure the motion of sources emitting visible and invisible kinds of electromagnetic radiation. (See also note 32 in this chapter.)

38. As we mentioned before, we measure only the radial component of the star's velocity, that is, the component of the velocity pointing toward our telescope.

39. David Charbonneau, Timothy M. Brown, David W. Latham, and Michel Mayor, "Detection of Planetary Transits Across a Sun-like Star," *Astrophysical Journal* 529, no. 1 (2000): L45–48.

40. For readers who want up-to-date results, see the list from NASA of exoplanets at https://exoplanets.nasa.gov/discovery/discoveries-dashboard/.

41. Mercury has a 3:2 resonance, which means it rotates about itself 1.5 times at every orbit about the Sun. Since Mercury's orbital period is 88 days, one "day" on Mercury corresponds to 176 days on Earth.

42. Incidentally, Earth 2.0 is the name of a planned Chinese mission to find planets with the same mass and radius as Earth's and orbiting G-type stars with an orbital period of one year—the closest we can get to our own planet with a focus on astronomical properties. The mission is supposed to launch in 2026.

CHAPTER FIVE
LIFE ON OTHER WORLDS

1. Arthur C. Clarke, "Hazards of Prophecy: The Failure of Imagination," in *Profiles of the Future: An Inquiry into the Limits of the Possible*, rev. ed. (New York: Harper & Row, 1973), 36.

2. Arthur C. Clarke, *2001: A Space Odyssey* (New York: New American Library, 1968), 227.

3. See, e.g., Erich von Däniken, *Chariots of the Gods* (New York: Berkley Books, 1998).

4. Carl Sagan, foreword to *The Space Gods Revealed: A Close Look at the Theories of Erich von Däniken*, by Ronald Story, 2nd ed. (New York: Barnes & Noble, 1980), xiii.

5. Clarke and Kubrick worked jointly on the novel, although only Clarke's name appears as author, possibly because Clarke had put forth many ideas used in the novel in a series of short stories he published dating back to the early 1950s.

6. For the curious: Spectral lines are related to the specific energy levels of the atoms or molecules. According to quantum physics, electrons can orbit the atomic nucleus only in specific orbits. As electrons "jump" between orbits, they will absorb (going up one or more orbits) or emit (going down one or more orbits) photons of light with energy equal to the difference in energy between orbits. For molecules, the wealth of

spectral lines is related to vibrational and rotational motions that are excited when photons are emitted or absorbed, which are also discrete, or quantized.

CHAPTER SIX
THE MYSTERY OF LIFE

1. Svante Arrhenius, *Worlds in the Making: The Evolution of the Universe* (New York: Harper & Row, 1908); I. S. Shklovskii and Carl Sagan, *Intelligent Life in the Universe* (New York: Dell, 1966); F. H. Crick and L. E. Orgel, "Directed Panspermia" *Icarus* 19, no. 3 (1973): 341–346.

2. The idea of the "world-bearing turtle" is said to originate in a Hindu myth, first mentioned in Europe at the closing of the sixteenth century in a letter by the Jesuit Emmanuel da Veiga: "Others hold that the earth has nine corners by which the heavens are supported. Another disagreeing from these would have the earth supported by seven elephants, and the elephants do not sink down because their feet are fixed on a tortoise. When asked who would fix the body of the tortoise, so that it would not collapse, he said that he did not know." It beautifully illustrates the notion of infinite regression, a causal link that has no end, since every step depends on another prior one. Even if it has been traditionally used to describe the nature of the Cosmos—What does the world stand on?—we can see how attempts to describe an abrupt event from no prior cause, such as the origin of the Universe, will fall into the same logical trap. For a twentieth-century citation, see Jarl Charpentier, "A Treatise on Hindu Cosmography from the Seventeenth Century (Brit. Mus. MS. Sloane 2748 A)," *Bulletin of the School of Oriental and African Studies* 3, no. 2 (1924): 317–342.

3. Adam Frank, Marcelo Gleiser, and Evan Thompson, *The Blind Spot: Why Science Cannot Ignore Human Experience* (Cambridge, MA: MIT Press, forthcoming).

4. Readers familiar with the interpretational challenges of quantum mechanics will no doubt recognize the parallel with the "shut up and calculate!" approach to quantum physics.

5. L. E. Orgel, *The Origins of Life: Molecules and Natural Selection* (London: Chapman & Hall, 1973); Robert Alberts, Alexander Johnson, Julian Lewis, Martin Raff, Keith Roberts, and Peter Walter, *The Molecular Biology of the Cell*, 5th ed. (New York: Garland Science, 2002).

6. Peter Ward and Joe Kirschvink, *A New History of Life: The Radical New Discoveries About the Origins and Evolution of Life on Earth* (New York: Bloomsbury, 2015), 35.

7. Notably, Gerald Joyce's experiments with his collaborators have greatly contributed to our current understanding of in vitro evolution at the molecular level: Katrina F. Tjhung, Maxim N. Shokhirev, David P. Horning, and Gerald F. Joyce, "An RNA Polymerase Ribozyme That Synthesizes Its Own Ancestor," *Proceedings of the National Academy of Sciences (PNAS)* 117, no. 6 (2020): 2906–2913, https://doi.org/10.1073/pnas.1914282117.

8. This article summarizing current research on the origins of life illustrates this point quite clearly: Adam Mann, "Making Headway with the Mysteries of Life's Origins, *PNAS* 118, no. 16 (2021): e2105383118, https://doi.org/10.1073/pnas.2105383118.

9. Carol Cleland and Christopher Chyba, "Does 'Life' Have a Definition?," in *The Nature of Life: Classical and Contemporary Perspectives from Philosophy and Science*, eds. Mark A. Bedau and Carol E. Cleland (Cambridge: Cambridge Univ. Press, 2010), 326.

10. Laboratory for Agnostic Biosignatures, https://www.agnosticbiosignatures.org/.

11. Paul Davies, *The Fifth Miracle: The Search for the Origin and Meaning of Life* (New York: Penguin, 1998), 260.

12. P. W. Anderson, "More Is Different: Broken Symmetry and the Nature of the Hierarchical Structure of Science," *Science* 117, no. 4047 (1972): 393–396, at 393.

13. Ernst Mayr, *This Is Biology: The Science of the Living World* (Cambridge, MA: Harvard Univ. Press, 1997), 37.

14. Stuart A. Kauffman, *Humanity in a Creative Universe* (Oxford: Oxford Univ. Press, 2016), 3.

15. Francis Bacon, *The Novum Organum*. See, e.g., SirBacon.org, http://www.sirbacon.org/links/4idols.htm.

16. Francisco J. Varela, "The Creative Circle: Sketches on the Natural History of Circularity," in *The Invented Reality*, ed. Paul Watzlavick (New York: W. W. Norton, 1984), 2, 3.

17. Scientists define terrestrial planets as worlds with a diameter between 0.5 and 1.5 that of Earth, while Sunlike stars have surface temperatures between 8,180°F and 10,880°F (4,527°C to 6,027°C). As discussed in part II, habitable zones are notoriously hard to define, being subject to many variations and local quirks. Venus, for example, is just outside the sun's habitable zone but has a very hard environment for sustaining life; the same is true for Mars, although life could have existed there billions of years ago or even persist subsurface. These local variations add complications to the definition of habitable zones, and to possible life-bearing worlds—those with subsurface oceans, like Europa, or those that had life in the distant past or that will have life in the future. The concept of the habitable zone is a blunt knife for thinking about life elsewhere, which is tailored to consider rocky worlds capable of bearing liquid surface water. See Steve Bryson, Michelle Kunimoto, Ravi K. Kopparapu et al., "The Occurrence of Rocky Habitable Zone Planets Around Solar-Like Stars from Kepler Data," preprint, submitted November 5, 2020, https://arxiv.org/pdf/2010.14812.pdf.

18. J. Richard Gott III, "Implications of the Copernican Principle for Our Future Prospects," *Nature* 363 (1993): 315–319.

19. Peter Ward and Donald Brownlee, *Rare Earth: Why Complex Life Is Uncommon in the Universe* (New York: Copernicus Books, 2000).

20. I use the word "possibly" to stress that there is no teleological endpoint for life's evolution, certainly none that points to intelligent multicellular life.

21. Elizabeth Kolbert, *The Sixth Extinction: An Unnatural History* (New York: Picador, 2015).

22. Peter Ward and Joe Kirschvink, *A New History of Life: The Radical New Discoveries About the Origins and Evolution of Life on Earth* (New York: Bloomsbury Press, 2015).

23. Marcelo Gleiser, *A Tear at the Edge of Creation: A Radical New Vision for Life in an Imperfect Universe* (New York: Free Press, 2010), ch. 53.

24. For the readers that justifiably don't remember their high school biology, prokaryotic cells don't have a distinct nucleus for their genetic material nor other specialized organelles, while eukaryotic cells (the ones we are made of) have genetic material such as DNA in the form of chromosomes contained inside a distinct nucleus.

25. Dorion Sagan and Lynn Margulis, *Microcosmos: Four Billion Years of Microbial Evolution* (Berkeley: Univ. of California Press, 1997).

26. Paul Davies, *The Eerie Silence: Renewing Our Search for Extraterrestrial Intelligence* (New York: Houghton Mifflin Harcourt, 2010).

CHAPTER SEVEN
LESSONS FROM A LIVING PLANET

1. The story of how these bacteria emerged and changed the planet is truly fascinating but not needed here for our purposes. I encourage the interested reader to explore Peter Ward and Joe Kirschvink, *A New History of Life: The Radical New Discoveries About the Origins and Evolution of Life on Earth* (New York: Bloomsbury, 2015), esp. ch. 5.

2. I tell this story in detail in my book *The Prophet and the Astronomer: A Scientific Journey to the End of Time* (New York: W. W. Norton, 2001).

3. Unless, of course, it was planted there by other intelligences, as we or our posthuman descendants might do if we ever populate Mars and other worlds.

4. Marcelo Gleiser, "From Cosmos to Intelligent Life: The Four Ages of Astrobiology," *International Journal of Astrobiology* 11, no. 4 (2012): 345–350.

5. Indeed, there is considerable confusion over the power of models to provide definitive solutions to scientific questions. They can be spectacularly successful when describing observed phenomena and can even be predictive of new, yet unobserved effects. The confusion starts when models are mistaken for the physical reality they aim to describe, what philosopher and mathematician Edmund Husserl called surreptitious substitution. Models are like maps to a territory: they simplify to be effective. But as with any map, scientific models need a conceptual structure upon which they are formulated. The experts would recognize this in, say, models of cosmological inflation that use a scalar field called the inflaton with interactions described by some specific potential energy. Where does the scalar field and its specific potential come from? Well, possibly from another layer of physical complexity underneath, such as superstrings. And where do superstrings come from? The usual answer is "they are fundamental," meaning that they don't come from anything else. But clearly there is no conceptual basis for making this kind of statement, given that superstrings themselves are formulated in a specific space-time and with a "string tension constant" that must come from somewhere. This "somewhere" fills the conceptual gap of the unknowable First Cause, from which physicists derive all models of cosmic origins in one way or another.

6. These were the nuclei of hydrogen, helium, and lithium, the first three chemical elements of the periodic table, with one, two, and three protons in their nuclei, respectively. Isotopes are variants of a chemical element with different numbers of neutrons in their nucleus. For example, deuterium is an isotope of hydrogen with a proton and a neutron in its nucleus, and helium-3 is an isotope of helium with two protons and one neutron in its nucleus.

7. For a thorough and highly engaging history of these developments, I recommend Jeremy DeSilva, *First Steps: How Upright Walking Made Us Human* (New York: HarperCollins, 2021).

8. As remarked before, defining which animals are capable of higher cognition (or even identifying the dividing line for higher cognition) is not very productive, given the imprecisions of our understanding and fossil record. Instead, I adopt a more pragmatic approach and link cognition at the level needed for my argument to the emergence of figurative art. We don't know precisely when that happened, but current evidence suggests that as far back as fifty-two thousand years ago, humans were representing aspects of reality through paintings. See, e.g., Maxime Aubert, Rustan Lebe, Adhi Agus Oktaviana et al., "Earliest Hunting Scene in Prehistoric Art," *Nature* 576 (2019): 442–445.

9. Barbara C. Sproul, *Primal Myths: Creation Myths Around the World* (New York: HarperCollins, 1991).

10. In my book *The Dancing Universe: From Creation Myths to the Big Bang* (New York: Dutton, 1997), I present a detailed analysis of creation myths from different cultures, contrasting them with modern cosmological models.

11. Einstein quoted in *The Quotable Einstein*, ed. Alice Calaprice (Princeton: Princeton Univ. Press, 1996), 158–159.

CHAPTER EIGHT
BIOCENTRISM

1. Recall that *anima* in Latin means soul.

2. Thomas Berry, *Evening Thoughts: Reflecting on Earth as a Sacred Community*, ed. Mary Evelyn Tucker (San Francisco: Sierra Club, 2006), 40.

CHAPTER NINE
A MANIFESTO FOR HUMANITY'S FUTURE

1. *Simone Weil: An Anthology*, ed. Siân Miles (New York: Grove Press, 2000).

INDEX

51 Pegasi, 126
2001 (Clarke), 136–37, 141, 224n5

abiogenesis, 151
absorption spectrum, 144
adaptability, 176, 178
air, 23, 28, 30, 80, 140, 173
alien life, 51, 95–98, 106–8, 116–21, 134, 135–40, 142, 152, 157
Alpha Centauri, 115, 136
amino acids, 151, 168
ammonia, 113, 114, 144, 183
Anaxagoras, 36
Anaximander, 27–32, 36, 41, 91
Anaximenes, 28, 36
Anderson, Philip, 158
Andromeda, 83
Anthropocene, 6, 167
Apeiron, 28–31, 33
Arcadia (Stoppard), 107
Aristarchus, 12
Aristotle, 12, 27, 29, 34, 35, 39, 42, 79
Arrhenius, Svante, 152
asteroid collisions, 5, 120, 167, 175, 177, 196

asteroids, 110
asters, 23
astrobiology, 152, 180–86
astronomers, 93–94, 99, 101–4, 109, 117, 123–24, 131–32, 144. *See also names of individual astronomers*
astrotheology, 137–38
Atomists, 32, 35–40, 43, 60, 64, 85–86. *See also names of individual Atomists*
atoms, 31, 41, 47, 61, 153, 154, 157, 222n32, 224n6
  and Atomists, 35–38, 42
  first, 180, 182
A-type stars, 118, 120, 223n35
Augustine of Hippo, Saint, 79
Australopithecus, 74

Bacon, Francis, 161
bacteria, 5, 105, 162, 164, 173, 175, 177, 228n1
Banks, Joseph, 92
Beethoven, Ludwig van, 90
Being vs. Becoming, 35–36
Bentley, Richard, 85–86

Berlin Observatory, 99
Berry, Thomas, 78, 191–92
Bible, 186
Big Bang, 1, 52, 95, 120, 183
bilateral symmetry, 105–6, 176
biocentrism, 7, 193–96, 202–7
biochemistry, 54, 153–56, 163, 169
Biological Age, 180, 183–85
biology, 64, 151, 152–53, 156–59, 164, 169–72, 228n24
biosphere, 4, 5, 6, 32, 54, 66, 115, 159, 162, 176
black holes, 59, 182
Blake, William, 94
Blind Spot, The (Frank and Thompson), 51
blue super giants, 117
Book Nobody Read, The (Gingerich), 16–17
Borges, Jorge Luis, 59
bottom-up forces, 155, 158, 162
bounce cosmological models, 43, 64
boundary conditions, 158
Brout, Robert, 100
brown dwarfs, 221n31
Brownlee, Donald, 167–68
Bruno, Giordano, 18, 80–81, 96, 200
B-type stars, 118, 119, 120, 223n35
Buys Ballot, C. H. D., 124

Calypso, 221n29
Cambrian explosion, 171, 176
carbon cycle, 175
carbon dioxide, 114, 145, 171, 175, 183
Cassini (space probe), 113
cataclysms, 5, 105, 156, 167, 175, 177, 184
Catholics, 82

causation, 39, 152, 158, 162, 186, 210, 229n5
cave paintings, 185–86
Ceres, 220–21n26
CERN, 50
Charbonneau, David, 131
Chauvet Cave, 185
Chemical Age, 180–81, 183
chemistry, 18, 64, 91, 112, 120, 127, 134, 144–45, 150–51, 156–58, 165, 169, 171, 179, 191, 229n6. *See also names of individual chemists*
chemotaxis, 162
Christianity, 17, 39, 75, 78–81
Clarke, Arthur C., 136–37, 141, 224n5
climate change, 6–7, 200–201
cognition, 185, 192, 230n8
Cognitive Age, 19, 180–81, 185–87
Coleridge, Samuel Taylor, 89, 92, 218n4
Colonial Theory, 171
cometary collisions, 5, 120, 167, 175
comets, 83, 91. *See also* cometary collisions
community, 73–77
comparative planetology, 55, 109
complex life, 67, 158, 167–68, 171–72, 184
Conard, Nicholas, 185
Constantine the Great, 77
constellations, 23
consumerism, 205–6
continent formation, 167, 171
Conversations on the Plurality of Worlds (Fontenelle), 97
Copernicanism, 3, 18–19, 38, 48, 52–55, 59, 64–65, 96–97, 111, 116, 179, 194

INDEX

Copernicus, Nicolaus, 2–3, 11–19, 38, 52–53, 80–81, 96, 111, 200
Cosmic Designer, 86
cosmic stories, 5–6
cosmological inflation, 229n5
Cosmos, 12, 16–18, 22–23, 24, 25–26, 28–30, 37–38, 39, 42, 45, 53, 56, 79, 80–84, 90–91, 94, 114, 116, 127, 130, 168, 172, 180, 200
Cosmos (TV series), 62
Counter-Reformation, 81
Creation, 6, 12, 17, 25, 26, 38, 80, 83, 87, 186, 191, 195. *See also* Creation stories
creation/destruction cycles, 32, 33–34, 37–38, 41–43, 55–56, 64, 114, 161
Creation stories, 20–21, 25–26, 186–87, 230n10
Crick, Francis, 152
cryovolcanoes, 113, 114
cyanobacteria, 173, 174–75
cyclic models, 64

Dante Alighieri, 17, 39, 80
dark energy, 56–57
dark matter, 56–57, 182
d'Arrest, Heinrich Louis, 99
Darwin, Charles, 18, 64, 94, 96, 151
Darwin, Erasmus, 94
da Veiga, Emmanuel, 225n2
Demiurge, 39, 86
Democritus, 32, 35, 36–38, 40, 42
Descartes, René, 53, 80
de Vlamingh, Willem, 98, 215n17
dialectics, 22
digital age, 4
Dimidium, 126

dinosaurs, 66–67, 159, 177–78
Divine Comedy (Dante), 17, 39
DNA, 153, 168, 171, 174, 228n24
Doppler, Christian, 124
Doppler effect, 124–26
Doppler method. *See* radial-velocity (or Doppler) method
dwarf planet, 108, 140, 220–21n26

Earth
  beginnings of life, 191, 222–23n35
  centrality of, 3, 16–18, 23–24, 26, 38, 53, 79–81
  change needed to preserve, 200–207
  cosmogonical model of, 29–30
  creation/destruction cycle of, 41–43
  diversity of, 55
  ever-changing nature of, 24
  exceptionalism of, 25–26
  gravity of, 44, 83, 84, 86
  materialistic view of, 3–4
  mechanical model of, 29–30
  mediocrity principle of, 53, 64–65, 96, 166
  objectification of, 2, 67
  orbit around Sun, 3, 12–13, 16–17, 38, 52–53, 65, 82–83, 86, 110, 133
  origin of life on, 5, 32, 33, 66, 68, 72, 105, 114, 120–21, 149–56, 168–81, 191–92, 193, 222–23n35
  as terrestrial planet, 227n17
  typicality of, 110–11, 115, 116, 127, 158
Earth 2.0, 134, 176, 224n42
Earthlike planets, 18, 38, 51, 65–66, 68, 128–29, 132, 134, 140, 143–45, 151, 166, 179

Einstein, Albert, 43, 45, 48, 68, 102, 187, 220n20
Einstein-Rosen bridges, 136–37
ekpyrosis ("out of fire"), 42–43, 64
electric fields, 46, 222n32
electricity, 45–47, 151
electromagnetism, 45–46, 214–15n13
    electromagnetic radiation, 46, 56, 182
    electromagnetic waves, 58, 222n32, 223n37
electrons, 32, 36, 58, 60, 158, 224–25n6
electroweak theory, 214–15n13
elementary particles, 36–37, 47, 49, 50, 56, 58, 157, 180
elements, four basic, 23, 33, 80
emotion, 24, 101, 103, 192, 202
Empedocles, 23, 32, 33–34, 43
empirical validation, 52, 63
Empyrean, 39, 80, 82
Enceladus, 113–14, 118
Englert, François, 100
Enlightenment, 53, 67, 87, 103, 202
entanglement of autonomies, 164–65
Enuma Elish (Babylonian "Epic of Creation"), 186
Epicurus, 32, 37–38, 42, 80, 85–86
epistemic humility, 57
epistemic nihilism, 57
E ring, 113–14
Escher, M. C., 163
ether, 42, 80, 102–3, 220n22
eubacteria, 175
eukaryotic cells, 169, 170–71, 174–75, 228n24
Europa, 112, 113, 114, 118, 227n17
evolution, 5, 18, 19, 32, 34, 64–65, 66, 68, 74, 96, 105–6, 140, 150–58, 165–87, 192–94, 199, 203, 209, 227n20
exceptionalism, 25–26, 195
exoplanets, 65–66, 74, 95, 101, 115–17, 121–22, 126–29, 131–34, 142–44, 172, 179, 223n36
extinction events, 167, 176
extrapolation, 52, 65–67, 98, 179
extraterrestrial life, 51, 95–98, 106–8, 116–21, 134, 135–40, 142, 152, 157
extremophiles, 119

failed stars, 221n31
faith-based belief, 22, 82
feedback loop, 156, 174
ferns, 175
fields, 43–46, 51, 109, 168, 220n20, 222n32. *See also* unified field theory
figurative art, 230n8
fire, 21–22, 159–60, 162
    Anaximander's mechanical model, 29–30
    basic element, 23, 33, 80
    manipulation of, 178, 184–85
    primal, 42–43
First Cause, 39, 152, 181, 186, 229n5
Fontenelle, Bernard Le Bovier de, 52, 96–97, 98–99
Frank, Adam, 51
Frankenstein (Shelley), 89–90, 151
French Academy of Sciences, 99, 102
Friedrich, Caspar David, 90
F-type stars, 118
fundamental forces, 43–48, 49, 157
fungi, 175

galaxies, 1, 5, 18, 32, 44, 50, 54–56, 58, 67, 68, 83, 109, 112, 113, 116, 119, 125, 127, 133–34, 135, 136, 138, 141, 143, 165–66, 179, 182–83, 191, 194

Galilei, Galileo, 17, 53, 58, 80, 81, 83, 97, 200, 218n8

Galileo (space probe), 113

Galle, Johann, 99, 219n17

gas planets, 55, 98, 109–10, 111, 126–27, 128, 131, 132–33, 143

gauge bosons, 214–15n13

Gell-Mann, Murray, 100

general theory of relativity, 43, 102

Genesis, 186

genetic mutations, 156, 175, 176, 177–78

genetics, 127, 153–54, 170, 228n24

geodesics, 45

geology, 154

geometry, 44, 60–61

George III, King of Great Britain and of Ireland, 93

Giese, Tiedemann, 11–15

Gingerich, Owen, 16–17

glaciation events, 175–76

gnomon, 22, 28

God, 26, 39, 45, 62, 78–79, 80, 82, 83, 85–87, 94, 105, 137, 186

God of the gaps argument, 62–63

gods, 20–22, 27–28, 34, 37, 39–41, 49, 77–78, 198, 209–10. *See also individual gods*

Goldilocks zone, 112

gravitational waves, 59, 100

gravity, 44–45, 56, 83–86, 87, 91, 99, 100, 102, 122, 157, 160–61, 182, 217n7, 220n20, 220–21n26

Great Oxygenation Event, 174

Greenblatt, Stephen, 41

G-type stars, 118, 119, 120, 128, 134, 143, 166

Guralnik, Gerald, 100

habitable zone, 65–66, 111–12, 117, 119, 129, 166, 221n27, 227n17

Hagen, Carl, 100

Halley comet, 83

HD 209458 b, 131

Heisenberg, Werner, 58

Helene, 221n29

Helios, 21

helium, 31, 47, 117, 182, 229n6

heredity, 168

Hermes, 27

Herschel, Caroline, 93

Herschel, William, 92–93, 95, 99, 113, 116

Higgs, Peter, 100

Higgs boson, 60, 100

high-energy physics, 37, 50

Hinduism, 21, 56, 225n2

Hipparchus, 218n5

Holocene extinction, 167

homeostasis, 162

hominin species, 74, 177, 181, 184

Homo, 171, 184–85

Homo erectus, 184–85

Homo sapiens, 1–2, 74–75, 184

host stars, 65–66, 109–10, 112, 120, 121–22, 126–33, 143–44, 179, 217n7

hot Jupiter, 126, 133, 143

Hubble, Edwin, 55

Hubble telescope, 125

hunter-gatherers, 2, 74–75, 94, 216–17n3

hurricanes, 159, 160, 162
Husserl, Edmund, 51, 229n5
Huygens, Christiaan, 96–97
hydrogen, 30–31, 47, 114, 117, 144, 182–83, 229n6

Indigenous cultures, 2, 27, 54, 74, 75–76, 82, 88, 90, 193, 198, 210
induction, 98, 111, 127, 215n17
Industrial Revolution, 88
infinity, 36, 37, 42, 87
infrared radiation, 95, 222n32
inorganic chemistry, 169
instruments, 58–60, 93–95
intelligent life, 5, 64–67, 104–5, 166, 169, 171–72, 178–79, 203
inter-being, 5, 207
interconnection of all things, 2, 6, 44, 76, 83, 145, 165, 175, 193–96, 203
in vitro evolution, 226n7
Io, 113
Island of Knowledge, The (Gleiser), 49

Jains, 187
James Webb Space Telescope, 95, 125
Jezero Crater, 107
Judaism, 82
Jupiter, 12, 53, 55, 56, 110, 112–13, 123, 126, 131, 132, 140–41, 142, 160, 221n31

Kauffman, Stuart, 159
Kepler, Johannes, 16, 17, 18, 53, 80, 81–82, 83, 96, 129–31, 185, 200, 219n12
Kepler mission, 128, 132
Keynes, John Maynard, 87

Kibble, Tom, 100
Kirk, G. S., 36
Kirschvink, Joe, 168
Krenak, Ailton, 76
K-type stars, 118
Kubrick, Stanley, 141
Kuiper belt, 108

Laboratory for Agnostic Biosignatures, 156
language, 2, 32, 43, 44, 48, 75, 181
large-numbers astronomy-based argument, 165–66
last universal common ancestor (LUCA), 175
law of action and reaction, 122–23
Lescarbault, Edmond, 101
LESS approach to sustainability, 205
Letter to Herodotus (Epicurus), 37
Leucippus, 32, 35, 36
Le Verrier, Urbain, 99–102, 219n17, 220n20
life
   alien, 51, 95–98, 106–8, 116–21, 134, 135–40, 142, 157
   as animated matter, 191–92
   biocentrism, 7, 193–96, 202–7
   chemistry responsible for, 91–92, 112, 114, 120, 151
   collective, 73–74, 165, 194–96, 207
   complex, 67, 158, 167–68, 171–72, 184
   conceptualizing, 154–59
   cosmic imperative, 18
   creation/destruction cycles, 30–31, 32, 33–34, 43, 56, 91, 114, 161
   Creation stories, 25–26, 186–87

definition of, 1, 149–50

evolution of, 5, 18, 19, 32, 34, 64–65, 66, 68, 74, 96, 105–6, 140, 150–58, 165–87, 192–94, 199, 203, 209, 227n20

extraterrestrial, 51, 95–98, 106–8, 116–21, 134, 135–40, 142, 157

habitable zone, 65–66, 111–12, 117, 119, 129, 166, 221n27, 227n17

intelligent, 5, 64–67, 104–5, 166, 168–69, 171–72, 178, 179, 203

interconnection, 2, 6, 44, 76, 83, 145, 165, 175, 193–96, 203

interplay between planet and, 173–79

mediocrity principle of, 53, 64–65, 96, 166

multicellular life, 169, 170–71, 176, 178, 184

origin of, 5, 32, 33, 66, 68, 72, 105, 114, 120–21, 149–56, 168–81, 191–92, 222–23n35

panspermia, 152

process of, 164

RNA world hypothesis, 153–54

simple, 42, 167–68

steps from no-life to intelligent life, 168–72

worlds not hospitable, 134

as worthless and expendable, 3–4

light, 46, 56, 60, 95, 100, 102, 124–26, 131, 144, 182, 220n22, 222n32, 223n37

lithium, 229n6

Little Prince, The (Saint-Exupéry), 113

Louis XIV, King of France, 77

Love, 33–34

Lowell, Percival, 104

LUCA, 175

Lucian of Samosata, 95–96

Lucretius, 40–41

luminosity, 119

lumpy balls, particles gathered into, 57

Luther, Martin, 15, 217n6

Macfarlane, Robert, 90

magnesian stone, 27

magnetic fields, 43–46, 109, 168, 222n32

magnetism, 43–44, 45–46

mantras, 186

Maori, 187

maps, 28, 51–52, 57, 59, 63, 75

Margulis, Lynn, 170

Mariner program, 106

Mars, 12, 55, 66, 98, 103–8, 109, 110, 140, 141, 142, 220n224, 227n17, 228n23

Mars (Lowell), 104

Mars Sample Return program, 108, 220n224

Martian rovers, 106–8, 140

mass, 66, 122–23, 126, 128, 131–32, 133, 217n7

Mathematical Principles of Natural Philosophy [Principia] (Newton), 84–85, 91, 97

mathematics, 99–101

matter

as continuum, 42

dark, 56–57, 182

divided to indivisible atoms, 42

elementary particles, 36–37, 47, 49, 50, 56, 58, 157, 180

interaction at high energies, 52

matter (*continued*)
  life as animated matter, 191–92
  living, 159, 168–72
  as machinelike things made of atoms, 3
  natural rhythms, 41–42
  nonliving, 159, 168–72
  ordinary, 56
  unified theory of, 33, 37
Mayor, Michel, 126
Mayr, Ernst, 158
measurements, 51, 58–59, 63, 122, 164–65
mediocrity principle, 53, 64–65, 96, 166
Mercury, 12, 66, 101, 102, 103, 110, 126, 129, 133, 221n28, 224n41
Messier, Charles, 93
metabolism, 153, 154, 174
methane, 114, 144, 145, 175, 183
methanogenesis, 114
microbiology, 94
Micromégas (Voltaire), 53–54
microorganisms, 114
Milky Way, 54, 55, 109, 115, 119, 165
Mimas, 114
MINDFUL approach to consumerism, 205–6
miracles, 79, 82, 216n21
mitochondria, 170, 174
models, 6, 15–16, 28–30, 32, 43, 50–52, 56, 64, 79–80, 152, 158, 177, 181–82, 216n22, 229n5, 230n10
monism, 33
Moon, 12, 29, 30, 44, 71, 72, 79, 80, 95, 97, 108, 110, 141, 219n12, 221n28
Moonlight Sonata (Beethoven), 90

moons, 30, 55, 109, 112–15, 121, 140–41, 165
moral hierarchy, 75–77
moral rules, 195
MORE approach to engagement with the natural world, 205
motion
  affects on sound waves, 124–25
  electromagnetism, 45–46
  gravitational field, 44–45
  measuring sources of electromagnetic radiation, 223n37
  planetary, 81–83, 129–30
  Primum Mobile, 39, 80
  third law of, 122–23
  Unmoved Mover, 39, 80
  wobbling, 101, 122–24, 126
Mountains of the Mind (Macfarlane), 90
Mount Everest, 106
Mount Olympus, 27
Mount Wilson, 55
M-type stars, 117, 118, 119–20, 128
Muir, John, 44
multicellular life, 169, 170–71, 176, 178, 184
multiverse, 31–32, 37, 42–64
My First Summer in the Sierra (Muir), 44
myths, 20–22, 25–26, 28, 186, 225n2, 230n10

NASA, 106, 107, 113, 128, 131, 141, 150, 156
natural philosophy, 84
natural selection, 64, 96, 105, 150, 156, 157–58, 167, 170
natural spirituality, 206–7

INDEX

natural world, 75, 78
Nature
   animistic, 27–28
   constants of, 62–63, 91
   emergent complex systems of, 158
   as enemy, 78
   fundamental forces of, 43–49
   hidden forces of, 27–28
   humans as apex, 25–26, 53
   models of, 51
   MORE approach to engagement with the natural world, 205
   myths, 21–22
   objectification of, 89
   observation of, 58
   plundering of, 2–3, 54, 88, 192–93, 197–99
   power to control, 21–22, 71
   as rational, 44
   rational divine presence of, 68
   reading book of, 16–17
   reassessment of place in, 7–8
   resacralization of, 208–10
   as sacred realm, 2, 24–25, 75, 79, 82, 90, 186–87
   scientific worldview of, 3–4
   string theory and, 60
   unification of forces of, 48, 49–50
Nature of Things, The (Lucretius), 40–41
Neanderthals, 74, 185
nebulae, 93, 95
Neptune, 53, 55, 101, 108, 110, 143, 219n13
Neptunelike planets, 133
Neptunes, 143

neutrinos, 47–48
neutrons, 47, 182, 229n6
New History of Life, A (Ward and Kirschvink), 168
Newton, Isaac, 44–45, 53, 80, 83–85, 91–92, 97, 99, 100, 102, 114, 122–23, 218n8
Nhat Hanh, Thich, 5
nitrogen, 151
nonequilibrium dissipative structures, 159
nonlife, 65, 66, 150, 152, 155, 158, 162
nonliving dissipative structures, 159, 160, 162
nuclear fusion, 47, 118, 160–61, 182, 221n31
nucleic acids, 168, 169
nucleotides, 168

Olympus Mons, 106
"On Exactitude in Science" (Borges), 59
open thermodynamical systems, 162
Opportunity (rover), 106
organic chemistry, simple, 169
organized religions, 77–78, 82, 195
Orgel, Leslie, 152
oscillons, 57
Osiander, Andreas, 15–16, 217n6
O-type stars, 117, 118, 119–20, 222–23n35
ouroboros, 173
oxygen, 140, 144, 145, 159–60, 171, 173–74, 175, 176, 184, 201
ozone, 145, 173, 201

panspermia, 152
Paradiso, 80

Parmenides, 35
particle physics, 37, 50, 94
particles, 36–37, 47, 49, 50, 52, 56–57, 58, 60–61, 94, 100
Pastoral Symphony [No. 6 in F major] (Beethoven), 90
Paul III, Pope, 12, 217n6
Perseverance (rover), 107, 220n224
phosphates, 168
phosphorus, 151
photons, 62, 100, 214–15n13, 222n32, 224–25n6
photosynthesis, 174, 175
Physical Age, 180, 181–83
physics, 18, 32, 36–37, 43–44, 48, 50–52, 58, 60–63, 64, 74, 80, 81, 83, 84, 94, 102, 123, 125, 127, 136, 143, 156, 164, 179, 191
planetes, 23
planets. *See also individual planets*
    biocentric view of, 194–96, 203–7
    Chemical Age, 183
    comparative planetology, 55, 109
    discovery by Pre-Socratics, 23–24
    Earthlike, 18, 38, 51, 65–66, 68, 128–29, 132, 134, 140, 143–45, 151, 166, 179
    exoplanets, 65–66, 74, 95, 101, 115–17, 121–22, 126–29, 131–34, 142–44, 172, 179, 223n36
    formation, 1, 12, 29–30, 32, 37
    gas planets, 55, 98, 109–10, 111, 126–27, 128, 131, 132–33, 143
    gravitational field, 44–45
    gravity, 91
    habitable zone, 65–66, 111–12, 117, 119, 129, 166, 221n27, 227n17
    interplay between life and, 173, 175
    measuring effects on stars, 122–34, 223n38
    Milky Way and, 165
    motion of, 81–83, 129–30
    orbiting stars, 55, 80, 97, 109–10, 115, 143–44, 165–66, 179
    orbiting Sun, 3, 12–13, 15–18, 38, 45, 52–53, 65, 80, 82–83, 86, 93, 96–99, 101–3, 108, 110, 112, 126–27, 130, 133, 141, 224n41
    rocky planets, 66, 67, 82, 98, 109–11, 128, 132, 143, 166–68
    solar system exploration era, 140–41
    systems, 126–28, 131
    terrestrial planets, 24, 54, 79, 133, 143, 227n17
    transit, 101–2, 129–32
Plato, 39, 44, 86, 116, 200
Plutarch, 29
Pluto, 108, 140
Polydeuces, 221n29
pragmatism, 4
prediction, 99–100, 159
pre-Socratics, 20, 22, 27, 33. *See names of individual philosophers*
primal substance, 27–30, 42
Primum Mobile, 39, 80
prokaryotic cells, 169–70, 174, 228n24
Prometheus, 21
Protagoras of Abdera, 116
proteins, 151, 168, 169
Protestants, 82
protocells, 168–70
protolanguage, 184
protons, 47, 182, 229n6

protozoa, 170
Pythagoras, 44, 48

quantitative laws, 82
quantum physics, 164, 224–25n6, 225n4
quarks, 36, 47, 100, 158
Queloz, Didier, 126

radial-velocity (or Doppler) method, 122–29, 131, 132–33, 134
radiation
    electromagnetic, 46, 56, 182
    geophysical properties protecting life, 168
    habitable zone, 112
    human susceptibility, 140
    infrared, 95, 222n32
    outputs, 110, 111, 117, 119, 133, 181
    ultraviolet, 58, 106, 117, 174, 222n32
radio waves, 125
radius, 66, 128–29, 131, 133–34, 143, 221n26, 224n42
randomness, 175
Rare Earth (Ward and Brownlee), 167–68
rational cosmology, 28–29
rational divine presence, 68
rationalism, 103
Raven, J. E., 36
reality amplifiers, 58, 94, 99
reason, 53, 61, 74, 89, 103, 111, 194
red dwarfs, 117, 119
reflecting telescopes, 93, 218n8
Reformation, 81
refracting telescopes, 218n8
Rembrandt, 40, 42

Renaissance, 41
Revolutions of the Heavenly Spheres, On the (Copernicus), 12–16, 80
Rheticus, Georg Joachim, 13–16
RNA, 153–54, 168–69
RNA world hypothesis, 153–54
rocky planets, 66, 67, 82, 98, 109–11, 128, 132, 143, 166–68
Romantics, 89–90, 103
Rubenstein, Mary-Jane, 37

Sagan, Carl, 62, 138, 152
Saint-Exupéry, Antoine de, 48, 113
Saturn, 13, 54, 93, 109, 110, 113–14, 141, 221nn28–29
Schiaparelli, Giovanni, 103
science
    beginnings of, 22
    cosmic stories, 5–6
    as human construction, 49–50
    limits of scientific knowledge, 50–59, 63–64, 65, 94, 107–8, 156–57
    mechanistic, 53
    methodology, 52, 84
    power of, 59
    spirituality in, 68
    worldview, 3–4
Search for Extraterrestrial Intelligence (SETI), 142–43
secular spirituality, 202, 204
self-transcendence, 4
SETI, 142–43
Shelley, Mary, 89–90
Shiva, 21, 43, 56
Shklovskii, Iosif, 152
simple life, 42, 167–68
single gauge symmetry group, 214n13

Sirius, 54
Sixth Extinction, 6
Socrates, 20, 200
solar eclipse, 102
solar system, 3, 53–55, 65, 71, 93, 94, 97–98, 106, 108–17, 123, 126, 127, 128, 135, 140–42, 183, 195
solar ultraviolet radiation, 174
Somnium (Kepler), 18, 96
space, 32, 38, 43, 45, 53, 55–56, 59, 60, 71, 103, 140–41
spectral lines, 144, 224–25n6
spectral signature, 144
Spinoza, Baruch, 68
Spirit (rover), 106
spirituality, 202, 204, 206–7
Spock, 103
sponges, 175
spontaneous compactification, 60
Standard Model of cosmology, 50, 52
Standard Model of particle physics, 50, 52, 56
stars. *See also* Sun
    Alpha Centauri, 115, 136
    A-type, 118, 120, 223n35
    B-type, 118, 119, 120, 223n35
    in Christian medieval Cosmos, 80
    constellations, 23
    creation/destruction cycles of, 30–31, 32, 55–56
    failed, 221n31
    formation of, 1, 6, 24, 37, 50, 180, 182–83, 190
    F-type, 118
    gravity and, 83, 91
    G-type, 118, 119, 120, 128, 134, 143, 166
    habitable zone, 65–66, 111–12, 117, 119, 129, 166, 221n27, 227n17
    host, 65–66, 109–10, 112, 120, 121–22, 126–33, 143–44, 179, 217n7
    K-type, 118
    measuring effects planets have on, 122–34, 223n38
    mechanical model, 29–30
    Milky Way, 54, 55, 109, 115, 119, 165
    M-type, 117, 118, 119–20, 128
    nebulae and, 93
    as nonequilibrium dissipative structures, 159–62
    as nuclear fusion engine, 47, 183
    nurseries, 31, 161, 183
    O-type, 117, 118, 119–20, 222–23n35
    planetary transit, 101–2
    planets orbiting, 55, 80, 97, 109–10, 115, 143–44, 165–66, 179
    Sirius, 54
    Sunlike stars, 126, 131, 166, 227n17
    temperature of, 95
    transit of, 11, 12–13, 16, 19
    types of, 117–20, 223n35
Star Trek (television series), 103, 138
Star Wars (film), 138, 172
statistical inference, 57
Stoics, 42–43, 60, 64, 200
Stoppard, Tom, 107
strange loops, 164–65
Strife, 33–34
string theory, 43, 60–64
strong nuclear force, 47, 214–15n13
Suez Canal, 103
Sun
    creation/destruction cycles of, 42

Creation stories, 25
distance of Alpha Centauri from, 115, 136
as energy source, 184
gravitational field of, 44–45, 86, 220n20
as G-type star, 119, 120, 128, 134, 143, 166
law of planetary motion and, 81–82
life cycle of, 161
mechanical model of, 29–30
as nuclear fusion engine, 47–48, 117
orbit around Earth, 12, 80
as ordinary star, 54, 97
planets orbiting, 3, 12–13, 15–18, 38, 45, 52–53, 65, 80, 82–83, 86, 93, 96–99, 101–3, 108, 110, 112, 126–27, 130, 133, 141, 224n41
radiation, 168
temperature of, 95, 133
transit of, 20, 23
wobbling motion of, 123
sundials, 22, 28
Sunlike stars, 126, 131, 166, 227n17
super Earths, 133, 143
supernatural world, 75, 186
surreptitious substitution, 51, 229n5
sustainability, 205
swans, 98, 215n17
swarms, 57
Swerve, The (Greenblatt), 41
symbolic thinking, 2, 75, 178, 181, 186

Tear at the Edge of Creation, A (Gleiser), 48
tectonic plate drift, 167, 171

telescopes, 22, 53, 55, 57, 58, 71, 81, 92, 93, 98, 104, 113, 115, 121, 122, 129, 131, 133, 142, 223n36
Hubble telescope, 125
James Webb Space Telescope, 95, 125
reflecting telescopes, 93, 218n8
refracting telescopes, 218n8
Telesto, 221n29
temperature, 27, 95, 117, 118, 119, 126, 140, 158, 160, 171, 175, 222n32
terrestrial animal species, 32
terrestrial life, 107, 157
terrestrial physics, 44, 83
terrestrial planets, 24, 54, 79, 133, 143, 227n17
TESS, 132–33
Tethys, 221n29
Thales of Miletus, 27
theoretical physics, 48
theory, 99–101
Thompson, Evan, 51
three-dimensional reality, 60–61
tidal heating, 113
time, 12, 24, 28–29, 32, 37, 41–42, 43, 52, 64, 90
Titan, 114, 221n28
top-down causation, 158, 162
Transiting Exoplanet Survey Satellite (TESS), 132–33
transit method, 128, 129–34, 144
typicality, 110–11, 115, 116, 127, 158

UAPs, 138
UFOs, 138–39
ultraviolet radiation, 58, 106, 117, 174, 222n32

uncertainty principle, 58
unicellular organisms, 170–71
unidentified aerial phenomena (UAPs), 138
unidentified flying objects (UFOs), 138–39
unification, 48, 49–50, 214–15n13
unified field theory, 43, 45–48, 214–15n13
Universe, 1, 3, 4–6, 13, 18–19, 28, 32, 38, 43, 44, 46–47, 52, 54, 55–64, 67–68, 72, 74, 82, 87, 97, 105, 136, 143, 152, 165, 172, 179–82, 187, 195
Unmoved Mover, 39, 80
Uranus, 53, 92–93, 99, 110, 218n5

values, 195
Varela, Francisco, 163–64
Vatican, 15
Vedic period, 186
Venus, 12, 66, 101, 102, 109, 110, 129, 227n17
Viking program, 106
volcanoes, 106, 113, 120, 141, 167, 175
Voltaire, 53–54
von Däniken, Erich, 138
Voyager 1 (space probe), 140
Voyager 2 (space probe), 140
Vulcan, 102–3, 220n20

Walker, Sara Imari, 143
Wanderer Above a Sea of Fog (Friedrich), 90
Ward, Peter, 167–68
War of the Worlds (Wells), 104–5
water, 65, 104, 145, 171
  basic element, 23, 33, 80
  on Enceladus, 113–14
  on Europa, 112
  on Mars, 106–7
  as primal material, 27
  and rocky planets, 166–67
wavelengths, 95, 124–26, 144, 222n32, 223n37
weak nuclear force, 47, 214–15n13
Wells, H. G., 104–5
Wigner, Eugene, 100
wobble, 101, 122–24, 126
Wordsworth, William, 89, 92, 218n4
world-bearing turtle, 225n2
worldviews
  Atomistic, 36–41
  Copernican, 3, 16–19, 53–54
  humans as apex of Nature, 26
  impact of telescope on, 58
  mythic, 21–22
  Newtonian, 83–84, 87
  post-Copernican, 7, 68, 151
  pre-Socratic, 20–37
  scientific, 4
  unification, 47–48
  wrongheaded, 12
wormholes, 136–37

Young Rembrandt as Democritus the Laughing Philosopher, The (Rembrandt), 40

Zeno of Citium, 42
Zeus, 21, 27

Marcelo Gleiser is the Appleton Professor of Natural Philosophy and professor of physics and astronomy at Dartmouth College, a world-renowned theoretical physicist, and public intellectual. He has authored hundreds of technical and thousands of nontechnical papers and essays, as well as six books in English that have been translated into seventeen languages. His popular writings explore the historical, religious, and philosophical roots of science, past and modern. Gleiser is a fellow of the American Physical Society, a recipient of the Presidential Faculty Fellows Award from the White House and the National Science Foundation, and founder and past director of the Institute for Cross-Disciplinary Engagement at Dartmouth. With Adam Frank, he cofounded NPR's 13.7 Cosmos and Culture blog, and currently writes weekly for BigThink.com. He is the 2019 Templeton Prize laureate, an honor he shares with Mother Teresa, Archbishop Desmond Tutu, the Dalai Lama, and scientists Freeman J. Dyson, Jane Goodall, Francis Collins, Sir Martin J. Rees, and Frank Wilczek. Gleiser lives in Hanover, New Hampshire, among trees and brooks, with his wife and two younger sons. He is an avid trail runner, a practice he pursues as a bonding with Nature and with the discipline of the devout.